The Control of Metabolism

The Control of Metabolism

Edited by John D. Sink

The Pennsylvania State University Press

University Park and London

Library of Congress Cataloging in Publication Data

Sink, John D
The control of metabolism

1. Metabolism. 2. Biological control systems.
I. Title. [DNLM: 1. Metabolism—Congresses. QU120 C764 1970]
QP171.S58 574.1'33 73-12935
ISBN O-271-01124-6

In memory of

Henry Prentiss Armsby

1853 - 1921

Director, Pennsylvania Agricultural Experiment Station, 1887-1907; Dean, School of Agriculture, 1895-1907; Director, Institute of Nutrition, 1907-1921; whose dynamic research interest in metabolism is still quite evident in our University.

Contents

List of Contributors

Ransom L. Baldwin, Jr., B.S., M.S., Ph.D.
Professor of Animal Science, Department of Animal Science, University of California, Davis, California 95616.

B. R. Baumgardt, B.S., M.S., Ph.D.
Professor of Animal Nutrition and Head of the Department of Animal Science, The Pennsylvania State University, University Park, Pennsylvania 16802.

Arthur L. Black, B.S., Ph.D.
Professor of Physiological Chemistry and Chairman of the Department of Physiological Sciences, School of Veterinary Medicine, University of California, Davis, California 95616.

John R. Brobeck, B.S., M.S., Ph.D., M.D., LL.D.
Herbert C. Rorer Professor in the Medical Sciences, School of Medicine, University of Pennsylvania, Philadelphia, Pennsylvania 19104.

Michael L. Bruss, B.S., D.V.M.
NIH Trainee in Clinical Pathology, Department of Physiological Sciences, School of Veterinary Medicine, University of California, Davis, California 95616.

John D. Corbit, III, B.A., Ph.D.
Professor of Psychology, Walter S. Hunter Laboratory of Psychology, Brown University, Providence, Rhode Island 02912.

Edgar B. Hale, B.S., M.S., Ph.D.
Professor of Animal Behavior, Departments of Biology and of Poultry Science, The Pennsylvania State University, University Park, Pennsylvania 16802.

Alfred E. Harper, B.S., M.S., Ph.D.
Professor of Biochemistry and Chairman of the Department of Nutritional Sciences, The University of Wisconsin, Madison, Wisconsin 53706.

Howard T. Milhorn, Jr., B.S., M.S., Ph.D.
Associate Professor of Physiology and Biophysics and Codirector of Biomedical Engineering, Department of Physiology and Biophysics, School of Medicine, The University of Mississippi Medical Center, Jackson, Mississippi 39216.

Stuart Patton, B.S., M.S., Ph.D.
Evan Pugh Research Professor of Agriculture, Department of Dairy Science, The Pennsylvania State University, University Park, Pennsylvania 16802.

J. Thomas Reid, B.S., M.S., Ph.D.
Professor and Head of the Animal Nutrition and Physiology Division, Department of Animal Science, Cornell University, Ithaca, New York 14850.

Nathan E. Smith, B.S., M.S., Ph.D.
Assistant Professor of Animal Science, Department of Animal Science, Cornell University, Ithaca, New York 14850.

Jay Tepperman, A.B., M.D.
Professor of Experimental Medicine, Department of Pharmacology, College of Medicine, State University of New York, Upstate Medical Center, Syracuse, New York 13210.

Preface

As any cursory glance at the literature shows, the ideas of regulation and control permeate the entire fields of biology, agriculture, and medicine. Since they arose at many times and on different occasions, it is not surprising that an integrated treatise of such a complex collection of information is lacking.

The aim of this work was to bring together and integrate our current knowledge on metabolic control processes. This book focuses on the control of metabolism at the various levels of biological and social organization—molecular, cellular, organismic, extraorganismic, and theoretical. Hence, it should be of interest to the student, the researcher, and the teacher in the many agricultural, biological, and medical sciences, as well as sociologists, behavioral scientists, and engineers interested in regulation and control in animals.

To "set the stage," an overview of control, regulation, and homeostasis is presented in chapter 1. Chapters 2, 3, 4, and 5 then discuss the molecular and cellular aspects of the control of energy, carbohydrate, fat, and protein metabolism, and the regulatory function of biological membranes. Moving to the organismic level, chapter 6 discusses food intake, energy balance, and homeostasis; chapter 7, energy metabolism and the whole animal; and chapter 8, the control of thermoregulatory behavior. Extraorganismic control, as exercised by social structure, is discussed in chapter 9. Some theoretical aspects of the biological control system are presented in chapter 10; and finally, chapter 11 is an attempt to look into the future of research on the control of metabolism.

This treatise was developed from a symposium held at The Pennsylvania State University and dedicated to Henry Prentiss Armsby, an early pioneer of research in animal metabolism. May I express the hope that the material presented here will inspire its readers with exciting new ideas and new resolve for their achievement—a fitting tribute to Dr. Armsby.

J. D. Sink
University Park, Pennsylvania

Biographical Note

Henry Prentiss Armsby
(1853-1921)

> He has left a monument such as few men can boast and his work will live on and continue to fruit in the lives of those who suceed him.——E. W. Allen

These words were written about Dr. Henry Prentiss Armsby shortly after his death on 19 October 1921. They succinctly characterize the life of this eminent scientist and educator whose pioneering work in energy metabolism provided the basis and stimulus for the current work in this field throughout the world.

Henry Prentiss Armsby was born in Northbridge, Massachusetts, 21 September 1853, the only child of Lewis and Mary Prentiss Armsby. At the age of fifteen, young Henry entered the Worcester County Free Institute of Industrial Science (later Worcester Polytechnic Institute) and was graduated in 1871 with a Bachelor of Science degree. He remained at Worcester for one year as an instructor in chemistry.

The next two years were spent in graduate study at the Sheffield Scientific School at Yale College, where in 1874 he received the Bachelor of Philosophy degree. It was here that Armsby became acquainted with Samuel W. Johnson, professor of agricultural chemistry and director of the Connecticut Agricultural Experiment Station, under whose inspirational leadership Armsby embarked upon his remarkable career of investigation. After completing his studies at Yale, Armsby taught natural sciences in the High School at Fitchburg, Massachusetts, for a one-year period. This was followed by a year of intense work and study in Leipzig, Germany. During this sojourn abroad Armsby came into contact with Gustav Kuhn, director of the foremost agricultural experiment station of Germany, at Mockern, near Leipzig. This fruitful year of research on digestion and energy metabolism made a lasting impression on Armsby and without doubt determined the field of his future work.

Upon his return to United States he accepted a position at Rutgers College, New Brunswick, New Jersey, where he taught chemistry during the year 1876-77. From New Brunswick he was called back to New Haven on the recommendation of his former mentor, Samuel Johnson, to be chemist at the newly organized Connecticut Agricultural Experiment

Station. The four years in this position were important and eventful for Armsby, crowded with advanced studies, experimentation, and writing. He received the degree of Doctor of Philosophy from Yale University in 1879. In 1880 Armsby published *Manual of Cattle Feeding.* The first book of its kind in the United States, it was received with widespread favorable comment; it brought Armsby into prominence and foreshadowed his entire career.

In 1881, Armsby accepted the position of vice-principal and professor of agricultural chemistry at the Storrs Agricultural School (now the University of Connecticut), where he remained for two years. In 1883, Armsby was called to The University of Wisconsin to become professor of agricultural chemistry and associate director of the agricultural experiment station.

Even though Armsby was comparatively young—thirty-four—his name was synonymous with leadership in agricultural science, and in 1887 he was called to The Pennsylvania State College to become director of the agricultural experiment station. Armsby came to Penn State in the vigor of early manhood with a wealth of training and experience which had come to no other young American chemist at that time. He realized the fundamental importance of the energy aspects of animal production, and wrote extensively on this subject during his ten years at Penn State, pointing out the problems to the scientists in the United States Department of Agriculture. Their interest in the subject and the confidence they had in Armsby led to a proposal made to the college that he undertake, in cooperation with the Bureau of Animal Industry, a series of investigations into the fundamental principles of animal nutrition. In the spring of 1898 an agreement was reached between the USDA and The Pennsylvania State College to conduct a highly scientific nutritional research program at Penn State under Armsby's directorship.

After spending some time in Europe in the summer of 1898 studying various existing apparatus for respiration calorimetric research, Armsby decided to adopt the general principle of the respiration calorimeter at Middletown, Connecticut, which had been devised by Atwater and Rosa for experiments on man. The many technical problems encountered in the construction of such an apparatus for use with cattle were successfully solved by Armsby and his associates, Professors Fries and Osmond. Thus, the new respiration calorimeter was built at Penn State. The experimentation which began in 1902 may be regarded as the realization of Armsby's life-long ambition to demonstrate the dependence of farm practice on sound scientific principles of chemistry, physiology, and energetics. Years of work that followed attracted much attention in this country and abroad. Visitors from other countries interested in both animal and human nutrition, including medical groups studying nutrition, came to see his unique apparatus, and to benefit from his mastery of the many aspects of energy metabolism and animal nutrition. When the work of The Pennsylvania State College was divided into schools in 1900, Armsby became dean of the School of Agriculture. He retained this office until 1904 when, at his own request, he stepped down so that he might devote his entire time to the experiment station.

As director of the experiment station for twenty years (1887-1907), Armsby found that much of his time and energy were required for practical investigations, administrative work, and innumerable public demands. He held firm in his conviction that the future of the nation's agriculture must rest on scientific research, but as the institution grew, along with demands from the public, the available time for deep scholarly scientific research shrank. Consequently Armsby resigned from this position in 1907, to become director of the Institute of Animal Nutrition, which was then organized as a special division within the School of Agriculture. The remaining fourteen years of Armsby's life were spent in work involving his famous respiratory calorimeter.

His book *The Principles of Animal Nutrition,* published in 1903, was the outstanding publication in agricultural science of that era and brought to light his deep scientific insight into the many problems in animal nutrition. Another textbook, *The Nutrition of Farm Animals,* published in 1917, incorporated his clear understanding of the chemistry and physiology of nutrition with a vast amount of practical knowledge, making this book an admirable basis for teaching and for research.

In the summer of 1908, representatives from thirteen experiment stations and the Office of Experiment Stations of the USDA met at Cornell University to discuss the desirability of forming a permanent organization of animal nutrition investigators. At a meeting in November 1908, a constitution was adopted by thirty-two charter members. The name chosen for the new organization was the American Society for Animal Nutrition (now the American Society of Animal Science). Armsby was elected president of the society for the first three years of its existence. Under his capable leadership, the new society prospered, and much of its members' research was to be based on the fundamental principles of physiology and nutrition that he elucidated.

The many honors which came to Armsby reflected the high esteem in which his work was held by his colleagues in this country and abroad. In 1904 The University of Wisconsin conferred upon him the degree of Doctor of Laws, and from Yale University in 1920 and from Worcester Polytechnic Institute in 1921 he received the degree of Doctor of Science. He was elected a member of the Royal Society of Arts of Great

Britain in 1911, and a foreign member of the Royal Academy of Agriculture of Sweden in 1912. His portrait hangs in the Saddle and Sirloin Club of Chicago, a symbol of the highest tribute from the American Society of Animal Science.

Perhaps the most fitting epitaph to characterize the life of Dr. Armsby is the one which he himself referred to while delivering a memorial address at the funeral of Dr. George W. Atherton, President of The Pennsylvania State College, who died in 1906. *Si monumentum quaeris, circumspice*—if thou seekest his monument, look about thee.

1

Control, Regulation, and Homeostasis

John R. Brobeck

In physiology courses "homeostasis" is a recurring theme that students may not appreciate, and may even come to dislike. Nevertheless, it does not show its age, and probably will be discussed by generations of biologists yet unborn. "Control" and "regulation" sound perhaps more contemporary, maybe because they imply some connection with systems analysis, now a fashionable preoccupation in many disciplines and professions. Although it is true that in physiology the words "control" and "regulation" are often interchanged, I wish to distinguish between them. *Control* as I shall use the word means, simply, management. It is the management of a process, usually the government of its rate. *Homeostasis* is the name Walter Bradford Cannon (1929) gave to the constancy of composition of the internal environment that Claude Bernard (1878) had discovered to be a requirement for free living. We know now that the composition of living systems in fact is not constant; their constituents are always in flux, moving either toward or away from some mean position. This dynamic state (Schoenheimer, 1942) is more interesting as well as more significant than the constancy as such. It is the control systems of the body that keep the deviations not far from the mean or "normal" value. In many instances, these control systems receive information from specialized cells that sense the value of a given constituent, and thus serve to guide the functioning of the controls. To this situation I shall assign the word *regulation*. It means that the value of some physiological variable is controlled at a relatively constant level because it can be detected by sensory cells.

There are in living systems many examples of constancy that do not depend upon detection. A simple and well-known example is the following: if an animal is given arbitrarily a fixed amount of food, its body weight will become constant after either a gain or loss, depending upon the amount of food provided. This leveling-off occurs because body weight is a function of food intake, and total heat loss is a function of body weight.

So the mass of the body will increase or decrease until total heat loss comes to equal the energy available in the food, when any further gain or loss of weight becomes impossible. This equalizing is not a regulation of body weight; it does involve, however, regulation of body temperature. It is in the adjustment of heat loss to energy supply that the weight becomes stable, inasmuch as higher animals do not have control systems that sacrifice a constant body temperature in order to achieve gain or loss of weight. This is not to say that body mass is not regulated; it may be. In the experiment just outlined such regulation was prevented by administration of a constant quantity of food.

Exchanges

Controls and regulations, and therefore homeostasis, depend fundamentally upon exchanges of substance or energy between the body and its environment or between organs or regions within the body. If no exchange is going on, there can be no control, regulation, or homeostasis. Although this point may seem self-evident, it is not always explicitly recognized. How often is the nature of the exchange considered, for example, in discussions of homeostasis of arterial blood pressure? A good example of proper usage, and an appropriate one for this occasion, is the classic 1925 volume, *The Animal as a Converter of Matter and Energy,* by Henry Prentiss Armsby and C. Robert Moulton. In the title of this book the word "converter" identifies the energy exchange of higher animals. The quantitative relations of some of these exchanges—oxygen, water, heat, and energy—are shown in Table 1-1. Within the human body there are about 1½ L of oxygen, 42 L of water, 990 Kcal of heat (this is the heat I shall lose when I die and my body comes to room temperature), and about 147,000 Kcal of energy. This last is my body as hamburger for a cannibal. In daily exchanges there are obvious differences among the four variables. For oxygen the reservoir is small and the exchange is relatively large, 360 L a day. For water the content is large, and the exchange is only about one fourteenth as great. For heat the daily exchange is about four times the content, but for energy the daily exchange again is a small part of the content.

Figure 1-1 illustrates components of energy exchange. The "food intake" column represents the several foodstuffs, carbohydrates, protein, and fat of a typical human diet. The "total output" column shows the energy loss, the conversion of food energy into some other form. As the fractional columns indicate, this energy loss is composed of the basal heat

TABLE 1-1. Estimated Physiological Exchanges for a 70-kg Man at Room Temperature of 20° C

Variable	*Body content*	*Daily exchange*
Oxygen (L)	1.5	360
Water (L)	42	3
Heat (Kcal)	990	2,800
Energy (Kcal)	146,500	2,800

loss, the specific dynamic action (SDA), the heat of muscular exercise, and the work output. Now if there is indeed a regulation of body substance so that it tends to remain constant, it means that there is control of each of these variables or perhaps of all of them together—and both on the intake and on the output side of the exchange, or else with intake responding to output or vice versa. Unfortunately, almost nothing is known about control of several of these quantities. We do know a good bit about control of basal heat loss; furthermore, the SDA was extensively studied at Penn State by Forbes, Swift, Kriss, and others (Forbes et al., 1923-35) in work that remains the most nearly definitive on this phenomenon. Yet no one knows, for example, whether there is some control of motor output. What are the factors that determine how active we are, how much we are willing to work, or how much we actually do work from day to day or from week to week? Since this is such a large, uncertain variable, it makes the control

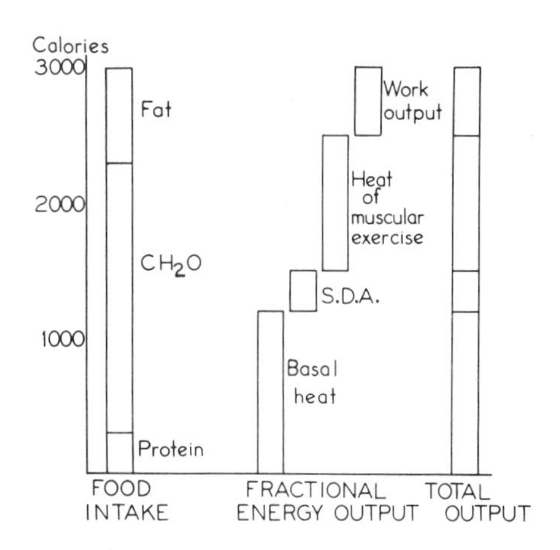

Figure 1-1. Comparison of energy intake, via typical human diet, with energy output and its several components.

of the length of this whole output column also uncertain. In young animals, the situation is further complicated by the phenomenon of growth. In spite of studies of growth hormone and its actions, and of hypothalamic releasing substances, how growth is ultimately controlled remains a mystery. The same can be said for the control of food intake.

One of the curiosities of control of energy exchange is that the two columns cannot be balanced by having food intake follow either total energy loss or total heat production. If either of these were to occur, the system would include a positive feedback loop that would destroy itself. This loop would be created by the SDA of food intake. When food is eaten, heat production and heat loss increase, as does total energy expenditure. If food intake were controlled so as to increase whenever heat loss became greater, the SDA would then call for more food, and would thereby evoke greater heat production, with more food, yet more heat, and so on, in what is called a "vicious circle." From this one can conclude that if food intake is linked to heat production, the feedback must be negative in sign. Thus, in all higher animals for which data are available, food intake decreases when the animal is subjected to any significant heat stress (Brobeck, 1960).

Control Mechanisms

Input, Output, and Content

The nature of physiological control mechanisms is illustrated in a very simple form in Figure 1-2. On the bottom line are the quantities, input, content, and output. Cannon and others have tended to focus their attention on the content and its apparent constancy, but that is only a part of the picture. One must examine the mechanisms by which input is converted into content, and content into output, and what mechanisms govern the rates of these two types of conversion. For any sequence like this the input-output relationship defines quantitatively the characteristics of the system.

Many available diagrams, especially in physiology, do not have clear and consistent meanings for arrows uniting the several terms. In this diagram, large arrows mean that some material or some identifiable quantity is converted from an input to a content and thence to an output. The thinner arrows indicate the pathways of nerve impulses, either to or from the central nervous system. On the afferent side they represent the possibility of detection, for example, of content, which might be detection of the water potential, the osmolarity, of fluid in the region of

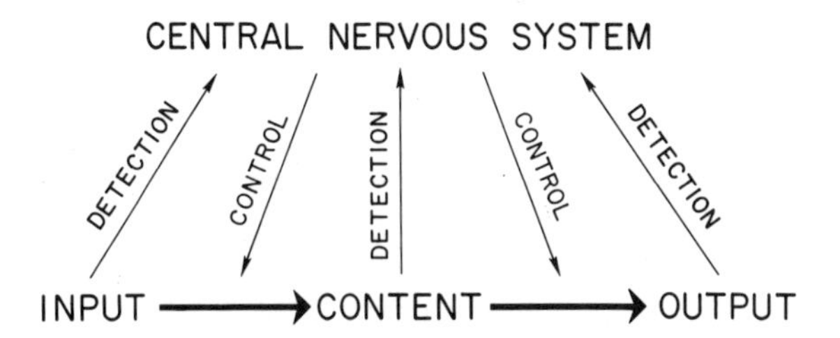

Figure 1-2. Simplified flow diagram for any biological exchange; detection is identified for some quality of input, of content, and of output. Input and output are controlled by the central nervous system. (Yamamoto and Brobeck, 1965)

hypothalamic osmodetectors. There is also a possibility of detection of something related to output, for instance, in temperature regulation, any change in skin temperature. Finally, there is the possibility of detection of something related to input; in temperature regulation this, too, might be a change in skin temperature if heat is being in the first instance gained via the surface of the body. The figure is intended to show that the central nervous system typically is made aware of input, content, or output, of changes in one or more of them, or of some combination of quantities or changes.

The control set of arrows illustrates that the nervous system is capable of controlling mechanisms that alter rate of input and/or rate of output, and hence, magnitude of content. If content is relatively constant, it is not because there is anything magic going back and forth between the nervous system and the content, as some discussions of homeostasis seem to imply. Rather, it is because the nervous system (perhaps via the endocrine system or in some other fashion) is able to control the size of the input and/or the output arrow. If content is found to be preserved constant, then intake and output must be controlled to approximately the same level.

Detection

To illustrate detection, I offer a new method of representing data. Consider a graph (Figure 1-3) on which the ordinate denotes some response of the body to some stated condition in the external environment. Then let the abscissa represent that environmental condition, such that the origin of the abscissa is a "poor" environment, while the right-hand end is a "good" or a "fit" environment, with reference to whatever variable may be

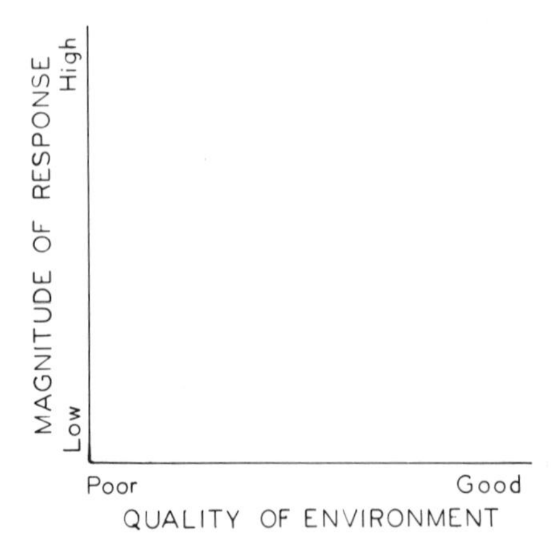

Figure 1-3. Axes for plotting some quality of the environment (abscissa) against the magnitude of the body's response to that quality (ordinate).

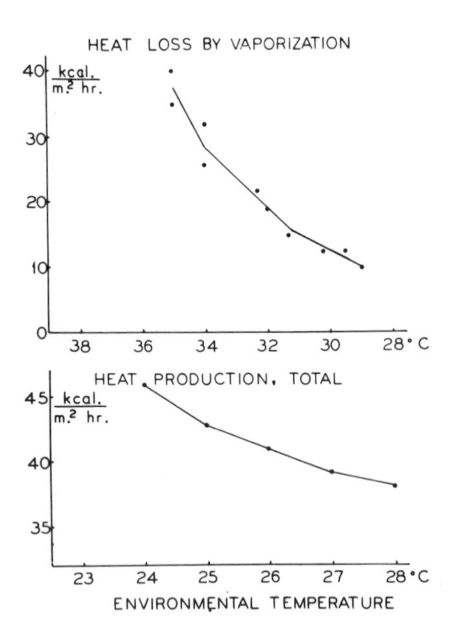

Figure 1-4. Top. When environment deviates from neutral temperature by becoming warm, body responds by sweating. (Crosbie et al., 1961). Bottom. If deviation is a lowered temperature, metabolic rate increases.

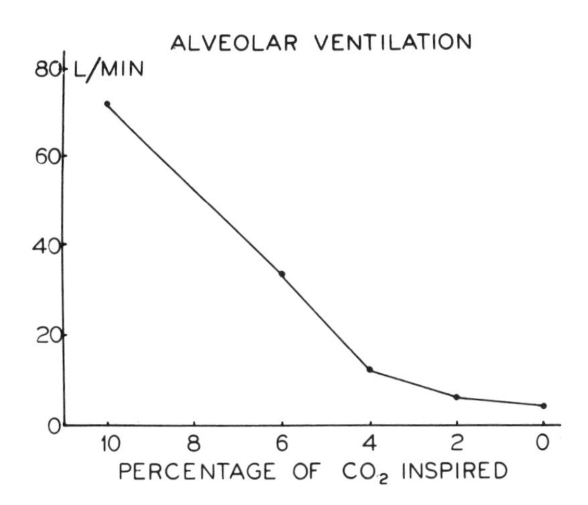

Figure 1-5. An increase in carbon dioxide content of inspired air evokes hyperventilation. (Yamamoto, 1960)

considered. The abscissa may represent coldness, heat, high CO_2, or low O_2—any one of these as a quality of a poor environment; then a neutral temperature, low CO_2 or normal atmospheric O_2 concentration is the good environment. In Figure 1-4 the upper ordinate is heat vaporized as sweat, whereas in the lower ordinate it is total heat production. Each of these is an appropriate response to an environmental change. The abscissa of each graph is temperature, but it is plotted in opposite directions above and below. In the top graph a hot temperature is not "good," but in the bottom graph it is a cold temperature that is not "good." Both abscissae have neutral temperatures at the right end. On a graph of this type even though the abscissa represents a variable in physical or chemical units, it portrays also the significance of the variable for the organism. Similarly, the ordinate depicts the magnitude of the organism's response, whether (in this figure) heat loss through sweating or heat production through shivering.

A similar relationship can be shown for the presence of CO_2 in the inspired air (Figure 1-5). As the concentration of CO_2 is increased and the environment becomes thereby less hospitable, the magnitude of the response is increased in the form of pulmonary hyperventilation. Environmental oxygen concentration yields a similar plot (Figure 1-6). In the lower curves, the CO_2 was allowed to vary as a result of the hyperventilation, but for the upper curve, the CO_2 concentration in alveolar air was arbitrarily held constant.

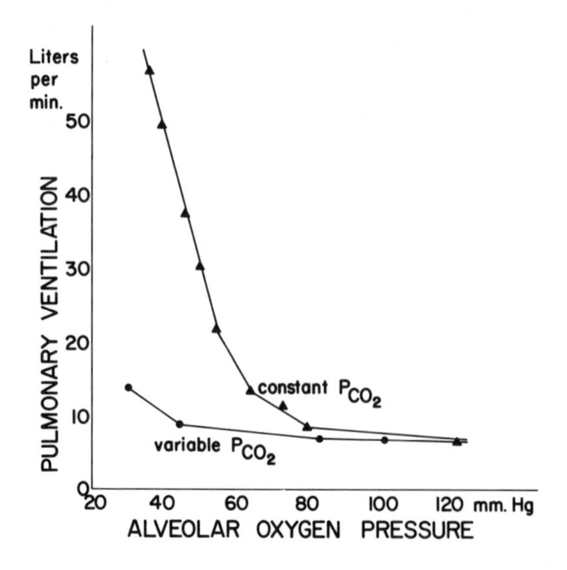

Figure 1-6. Lowering of ambient oxygen concentration increases ventilation, particularly when P_{CO_2} is held at a constant level. (Replotted from data of Loeschcke and Gertz, (1958), Fig. 4.)

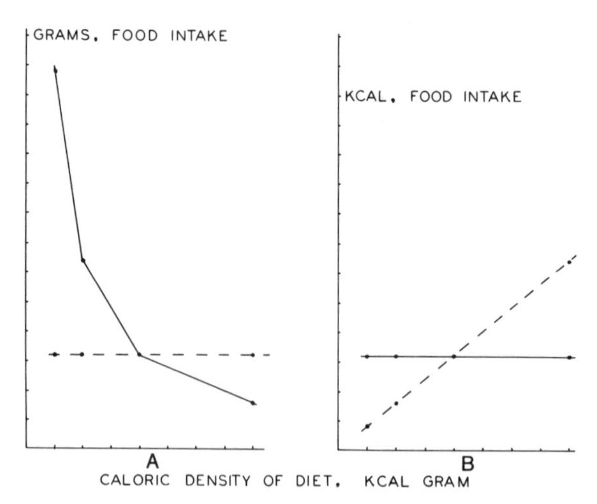

Figure 1-7. Food intake in grams (A) and kilocalories (B) when caloric density of diet is changed. Solid lines indicate constant energy intake; broken lines indicate constant intake in grams.

Is there a similar relationship for food intake? Graph A of Figure 1-7 is like the others. The abscissa represents the quality of the food available in the environment—i.e., the caloric value of the diet. At its end (on the right) is what is called a "rich" diet, with a high caloric density. For farm animals it would be a grain diet; in the laboratory it is a diet that has been loaded with fat. At its origin (on the left) is a low-density diet; we have used cellulose or mineral oil to make these diets, and have made mixtures with no food value at all. When farmers are obliged to give their animals only straw to eat, that of course would be a low-density diet. If an animal responds to the quality of the diet in the same way it responds to heat, cold, CO_2, or low oxygen levels, one would expect to see a relationship like the one of the first graph, the solid line, with an increase in total intake as caloric value of the diet is decreased. If such is what would happen, then the intake in kilocalories would remain constant (the solid line on Graph B). The animal would be controlling its food intake so that energy intake was constant, and increasing gram intake so as to achieve this constancy. On the other hand, if an animal is oblivious to changes in caloric density, there would be no change in food intake as caloric value is altered (broken line, Graph A), and intake of energy would show the simple proportionality illustrated by the broken line on Graph B. In this instance caloric intake is directly proportional to caloric density.

Owen Maller and I have performed experiments (unpublished) using rats given, for 24 hours, diets of compositions as specified in Figure 1-8. The results showed that the gram intake was almost constant, and that the kilocalorie intake was almost a linear function of the caloric density of the diet. We interpret these data to mean that a rat does not have a detector for the kilocalories in the diet, and consequently, it cannot show a conventional type of regulatory response in a 24-hour test in which caloric density is varied. It is true, of course, that animals given diets similar to these over a longer period of time do increase their gram intake of low-density mixtures. With the diets we were using, it took almost three weeks for the rats to increase their gram intake to a degree such that kilocalorie intake was the same as it was on the control diet. In the process of making this adjustment, however, they lost a significant amount of body weight. We believe that the enhanced gram intake was a response to some component or derivative of the weight loss and inanition, rather than to the low-density diet per se.

One should note also that if the material used to change the caloric density of a diet is water, the animals will show the predicted curvilinear relationship of gram intake against caloric density even in a 24-hour test (Figure 1-9). That is to say, rats respond to a dilution of calories with

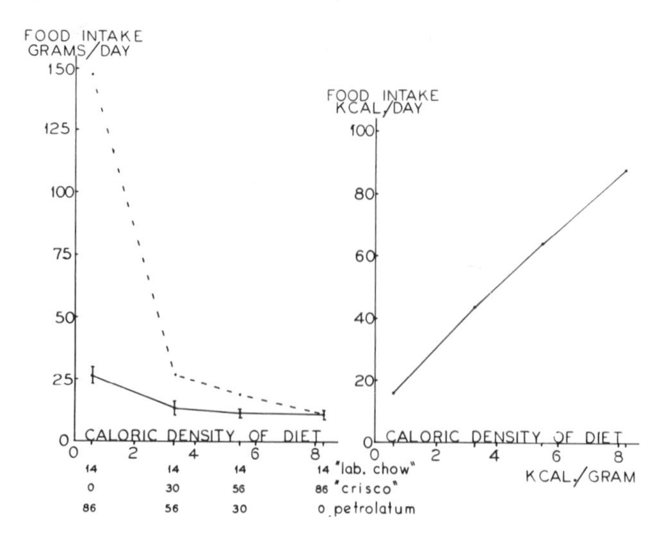

Figure 1-8. Experimental data when caloric density of diet was altered. Gram intake remained almost constant, energy intake was a linear function of caloric density. Broken line (left graph) denotes gram intake if kilocalorie intake had remained constant.

water in the same way the body responds to dilution of air with nitrogen, or with CO_2, or to a change in the environmental temperature. Yet if the dilution is made by adding or subtracting fat, with use of inert materials such as cellulose or mineral oil, the rats cannot recognize the change in the 24-hour period.

These experiments seem to show that animals have two possible sites of detection of changes. There is detection in the depth of the body that signals, for example, changes in partial pressure of oxygen or of carbon dioxide in arterial blood. The long-term response to inanition when poor-quality diets are fed probably comes from some sort of detector yet unknown but inside the body somewhere. By contrast, there is the possibility of detectors on the surface, as for example in thermal change, which responds to receptors in skin and mucous membranes. Although various investigators have concluded that the caloric value of food is somehow sensed as food is eaten, our data do not confirm such a conclusion. When a rat is in the process of eating, he has no way to find out whether the food is dense or dilute. It seems to me that perhaps the marketing of food for human use is based in part upon this quirk of physiology. This is why foods must be properly labeled. As a housewife goes along a supermarket aisle she may not know whether she is buying a food or a

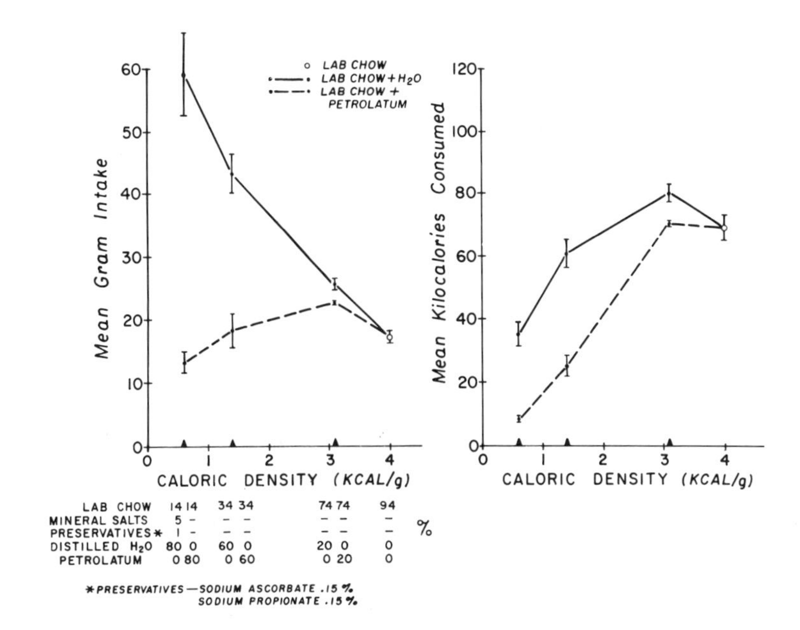

Figure 1-9. Experiments in which caloric density of diet was altered by addition of water (solid curves) or mineral oil (broken curves). Kilocalorie intake was more nearly constant when water was added than when mineral oil was used.

food placebo, and I suppose that in many cases she is not concerned about the difference. I have been told that in human nutrition every cellulose-containing foodstuff is to a certain degree a food placebo.

Models of Control Systems

Many models of control systems for physiological exchanges have been invented. We cannot consider all of them here. To my mind, however, Figure 1-10 is the most interesting—the proper word is "delightful"—model that has been proposed in the field of metabolism and energy exchange. It is Max Kleiber's model of a cow (1961), a model entirely consistent with all known physiological data, yet based upon hydraulic flow and pressure. Food intake is represented as controlled by a valve in the top funnel, with the height of this valve determined by a pulley system that is connected to a function of output, heat loss. The flow of water down through the system is also altered whenever energy is diverted

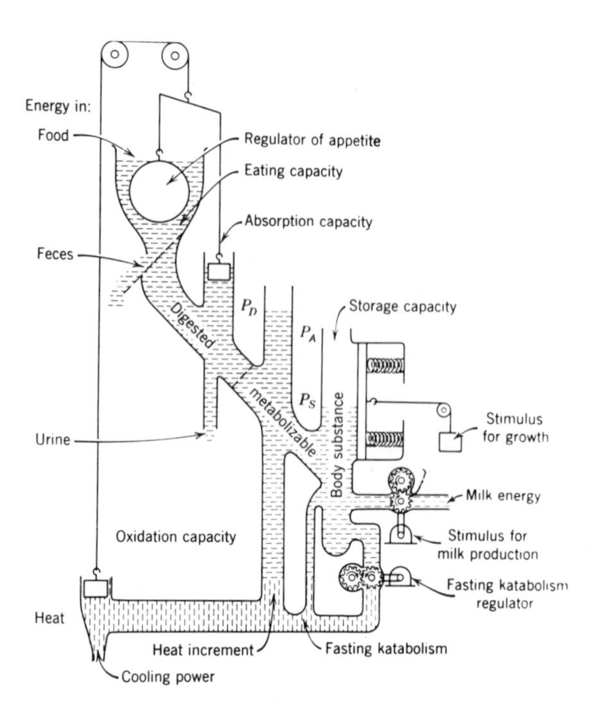

Figure 1-10. Kleiber's diagram (1961) of mechanisms that control food intake of a cow, utilizing a hydraulic model.

to the production of milk, or some other avenue of energy expenditure. The connection between the main valve and the outlet denoting heat loss is the negative feedback mentioned earlier; as heat production increases a plunger is lifted and the flow of energy into the system is cut off by a lowering of the main valve. The entire diagram may look a bit like a Rube Goldberg cartoon; but actually it is a penetrating analysis of how this control system works. One could perhaps write equations for this model derived from hydraulic flow and pressure, plus the dimensions of the tubes and chambers.

By contrast, Figure 1-11 is a diagram of a baby. It is not as amusing as Kleiber's, although it attempts to show specifically the mechanisms by which food intake is controlled. The central nervous system is shown merely as a "black box" that contains the neural control systems. Each arrow has the significance, "This process leads to the next process." Arrows returning at the top represent positive feedback with enhanced feeding, whereas those at the bottom are negative in sign.

If a mother places a hand on a baby's cheek, he will turn his head toward that side; in nursing the head is turned when the cheek touches the mother's breast. When something is placed between the baby's lips, he will grasp it and begin to suck. So feeding reflexes begin with touch sensation, and the sucking that is induced by touch in turn fills the mouth with milk, and leads to taste sensation that reinforces the grasping and sucking. (For simplicity the phenomenon of milk ejection is not included in this diagram.) These are the sources of positive feedback. As to the negative feedbacks, I suppose that they have not been studied actually in human infants. But from other experiments, many of them upon laboratory animals, there is believed to be some kind of metering system in the mouth and pharynx that signals when material has passed through it. There probably is a negative feedback from stretch of the stomach; at least, everyone knows the uncomfortable feeling of an over-full stomach. There is the possibility of osmodetection as a result of withdrawal of fluid from other compartments in the body, and also of chemodetection from a rise in blood sugar, change in amino acid level, or something similar. When the energy is passed on to the processes of metabolism, there may be thermal detection arising from changes in temperature regulation.

It is conceivable that each stage of this diagram could be amplified into a similar diagram of its own. For example, the stomach probably has a "preferred" degree of stretching. When it is not stretched that much, it sends out one kind of signal that we interpret as a hunger pain. When it is stretched too much, it sends another signal that is interpreted as satiety. If one could keep the stomach at a certain stage of stretch, it would never signal either hunger or satiety. It is possible to feed a baby with this result. A mother can care for a baby, if she does not have much

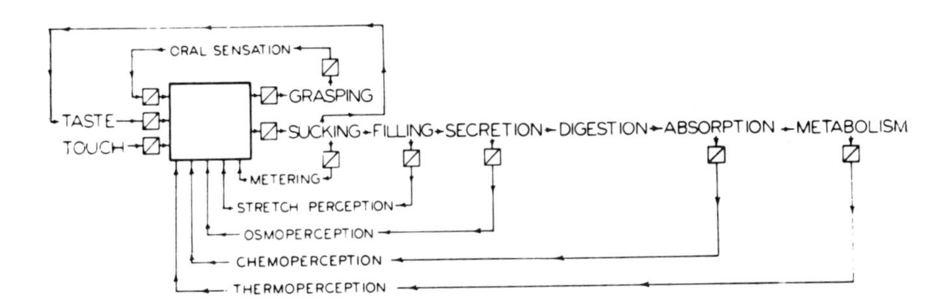

Figure 1-11. Control systems for nursing by human infant, with positive feedback loops at top, negative loops at bottom. Five types of detection are proposed in the negative loops.

else to do, throughout a 24-hour period without crying or other evidence of the baby's hunger, by merely nursing it each time before it has reached a stage of hunger. Yet the control system for the stomach is more complicated than only the sensory side, inasmuch as the rate of emptying is a function of the amount of material within the organ—which could also be included in the diagram. In a similar fashion, a more detailed diagram might be constructed for any level in body fluids that is presumed to contribute to chemodetection, osmodetection, or thermodetection.

Summary and Conclusions

Finally, I wish to suggest the possibility that models of physiological control systems may have some power, in the mathematical sense of the word, outside the very limited fields to which I have been applying these concepts. Exchange with control of input and of output seems to be a generalized situation in biology. When Yamamoto and I were preparing *Physiological Controls and Regulations* (1965), he distinguished between physiological and behavioral controls. He said that behavioral controls are not homeostatic, they are poikilostatic. In using behavior as a means of control and hence of regulation, the animal was said to make certain sacrifices of freedom in order to achieve the apparent constancy. I confess that this distinction is not clear to me. If one considers respiratory exchange (pulmonary ventilation with exchange of oxygen and carbon dioxide) as one example, and the keeping of the body at a proper wetness (not too wet, not too dry) as a second example, and keeping the body with proper energy reserves (not too fat, nor yet too lean), and with a proper reproductive pattern (not too many offspring, not too few), one proceeds from phenomena that we think of as automatic and therefore physiological to those that we consider to be largely behavioral. Yet there is no sharp boundary between the physiology and the behavior. Even pulmonary ventilation may require behavior, as, for example, when a person is forcibly held under water for 30 seconds.

So I conclude with the suggestion that there is a continuum of these phenomena, and that this continuum may even go well beyond physiology. It may extend into sociology, economics, and politics, certainly into the multitude of situations that have been fancied to be homeostatic in character. Wherever an exchange can be identified, with controls that determine the rate of input and/or of output, a physiological model may be the one that gives the greatest understanding of that system.

References

Armsby, H. P., and Moulton, C. R., 1925. *The Animal as a Converter of Matter and Energy*. The Chemical Catalog Company, Inc., New York.

Bernard, C., 1878. *Leçons sur les Phénomènes de la Vie Communs aux Animaux et aux Vegetaux*. Vol. 1, Ballière, Paris, pp. 67, 111-114, 123-124.

Brobeck, J. R., 1960. Food and temperature. *Recent Prog. Hormone Res.* 16:439.

Cannon, W. B., 1929. Organization for physiological homeostasis. *Physiol. Rev.* 9:399.

Crosbie, R. J., Hardy, J. D., and Fessenden, E., 1961. Electrical analog simulation of temperature regulation in man. *IRE Trans. on Bio-Medical Electronics* BME-8:245.

Forbes, E. B., et al., 1923-35. *Collected Papers on Investigations at the Institute of Animal Nutrition of The Pennsylvania State College*. The Pennsylvania State College, State College.

Kleiber, M., 1961. *The Fire of Life*. John Wiley and Sons, Inc., New York.

Loeschcke, H. H., and Gertz, K. H., 1958. Einflub des O_2-druckes in der Einatmungsluft auf die Atemtatigkeit des Menschen, gepruft unter Konstanthaltung des alveolaren CO_2-druckes. *Pflugers Arch. Physiol.* 267:460.

Schoenheimer, R. 1942. *The Dynamic State of Body Constituents*. Harvard University Press, Cambridge, Massachusetts.

Yamamoto, W. S., 1960. Mathematical analysis of the time course of alveolar CO_2. *J. Appl. Physiol.* 15:215.

Yamamoto, W. S., and Brobeck, J. R., 1965. *Physiological Controls and Regulations*. W. B. Saunders Company, Philadelphia.

2

Molecular Control of Energy Metabolism

Ransom L. Baldwin, Jr., and Nathan E. Smith

Early workers in nutritional energetics, in defining the basic terms "gross energy," "metabolizable energy," "net energy," and "heat increment" (specific dynamic action, work of digestion, thermal energy), attempted to select criteria and measurements for evaluations of feedstuffs and animal functions which could be related to specific aspects of physiological function and energy metabolism (Armsby, 1917). Metabolizable energy (ME) was defined as the gross energy of the feed minus the energy in feces, urine, and combustible gases, and was considered a measure of the energy from a feedstuff available for transformation in the body. Relevant to metabolizable energy and the necessity for defining additional terms to describe energy metabolism, Armsby (1917) stated: "The value of any nutrient or feeding stuff as a source of energy for maintenance is obviously measured by the extent to which it can diminish the loss of energy which the body would otherwise suffer," and "It was natural to suppose, therefore, that the metabolizable energy of a substance would represent its value for maintenance and this was long believed to be true, but later investigations have shown that such is not the case." The net energy value for maintenance (NE_m) of a feed was, therefore, defined as equivalent to the loss of body energy it prevents.

The difference between metabolizable energy and net energy for maintenance ($NE_m = ME - HI_m$) has been variously referred to as heat increment (HI_m), specific dynamic action, work of digestion, and thermal energy, and is considered a collective term reflecting the energy costs of digestion and assimilation of a feed (Armsby, 1917). In considering the utilization of feedstuffs for productive processes, early workers in nutritional energetics recognized that the efficiencies of utilization of metabolizable energy ($100 \times NE/ME$) for such varying processes as maintenance, growth, fattening, and lactation might vary (Armsby, 1917). Net energy values of feedstuffs for growth (NE_g) and lactation (NE_l) were defined as the portions of feed energy supplied in excess of an animal's maintenance energy requirement, appearing as energy in gain or milk, respectively. The collective terms, herein called heat increments of production and representing the differences between metabolizable energy values and net energy values of feedstuffs for growth, fattening, and lactation, were presumed to reflect costs of digestion, assimilation, and utilization of nutrients for productive processes.

Intermediary Metabolism

Armsby (1917) frequently speculated about the nature of the "vital processes" for which body energy and/or feedstuff NE_m were utilized, about the nature of the energy costs of digestion and assimilation (HI_m), and about other processes. In the course of discussing these and possible relationships between studies of intermediary metabolism and animal energy transformations assessed in balance studies, Armsby (1917) made the following comment: "The truth is that both types of investigation are equally necessary and each aids in the interpretation of the other. The balance experiment has been especially prominent in the past, while at present attention is being directed to a greater extent to investigations of the intermediary metabolism but neither can say to the other 'I have no need of thee.' " It appears to have taken quite some time for our knowledge of intermediary metabolism in animals to advance to the stage where we can begin to effect the union of nutritional energetics and intermediary metabolism visualized by early workers in nutritional energetics and suggested by Armsby.

Several major advances in our knowledge of intermediary metabolism were necessary before attempts to identify specific metabolic processes with nutritional energetic terms could be initiated (Blaxter, 1962; Krebs, 1964; Baldwin, 1968). First, we needed detailed information of metabolic pathways and physiological processes leading to conversions of nutrients to metabolic products and heat energy. Another essential advance was the identification of ATP (adenosine triphosphate) and recognition of the fact that the high-energy phosphate bonds (~P) of ATP provide a central or common carrier mechanism for the transfer of chemical energy released during transformations and oxidations of nutrients to sites where chemical energy is utilized in the performance of maintenance and productive functions. This observation enables convenient evaluations of stoichiometric relationships between catabolic, ~P-yielding functions, and anabolic, ~P-requiring and -using functions. The data summarized in Table 2-1 represent an evaluation of the heat energy equivalents of high-energy phosphate bonds of ATP (~P) formed during the oxidation of several common energy sources and utilized for

Table 2-1. Estimation of Heat Energy Equivalent of High-Energy Phosphate Bonds of ATP Formed from Common Energy Sources and Utilized in Maintenance Functions

Energy source	*~ P Bonds formed (~ P/mole)*	*ΔH_c (Kcal/mole)*	*ΔH_c/~ P (Kcal/bond)*
Glucose	38	673	17.7
Stearate	146	2712	18.6
Acetate	10	209	20.9
Propionate	18-19	367	20.4
Butyrate	27	524	19.4

maintenance functions. Use of the term "maintenance" in this case implies that initial and final states in the system are the same, except that ~P bonds were formed from an energy source provided and hydrolyzed in maintaining or regaining the initial state. Values for heat release per ~P, bond formed and utilized vary between 17.7 Kcal for glucose and 20.9 Kcal for acetate (Table 2-1). For mixed substrates, a generalized value of 19 Kcal per ~P was selected for use in formulations presented later in this chapter, unless otherwise indicated.

Currently it appears that most pathways of intermediary metabolism in animals have been identified and characterized. However, required stoichiometric relationships for several important types of metabolic processes, including, for example, the number of ~P bonds required in protein synthesis, and in ion and metabolite transport, have not been fully substantiated (Baldwin, 1968; Caldwell, 1966; Levin, 1969). The numerous, though yet incomplete, data derived from in-vivo and in-vitro radioisotope tracer studies (Annison et al., 1967; Black, 1970; Baldwin and Smith, 1971a) and from in-vivo arterial/venous blood metabolite difference or tissue metabolite uptake studies (Linzell, 1967) are essential to attempts to relate intermediary metabolism to nutritional energetics. Quantitative data of these types are necessary in establishing patterns and rates of nutrient assimilation and utilization, and defining precursor-product relationships. Another very important type of knowledge required in analyses of animal energetics is information on cellular, tissue, and animal metabolic regulatory systems. It must be recognized that animal intermediary metabolism is a dynamic, continuously changing process. Regulatory systems acting at cellular, tissue, and animal levels determine the nature of metabolic changes that occur after eating, during the transition from fed to fasting metabolism, after changes in diet and changes in physiological state, and in support of production. Because of the dynamic nature of animal metabolic processes and the complexities of metabolism and metabolic regulation, it has been suggested that computer-assisted simulation modeling techniques must be utilized in relating detailed aspects of intermediary metabolism to nutritional energetics (Baldwin and Smith, 1971a). However, if the justifiable simplifying assumption is made that ~P utilization, for essential maintenance and productive functions, controls and determines rates of nutrient utilization for ~P formation (Krebs and Kornberg, 1957; Mahler and Cordes, 1966; Garfinkel et al., 1968; Jobsis, 1964; Defares, 1964), instructive analyses of energy costs of many physiological and metabolic functions can be undertaken (Baldwin, 1968). These yield insight in identifying and evaluating the significances of intermediary processes in terms of the energy balance studies of nutritional energetics. Intermediary aspects of energy transformations associated with maintenance, heat increments of feeds fed at maintenance levels, and heat increments of feeds fed for production above maintenance levels will be analyzed and discussed below.

Maintenance Functions

In considering the nature of the intermediary metabolic processes leading to the energy expenditures of maintenance, a number of approaches can

Table 2-2. Estimated Energy Expenditures in Several Major Tissues

Tissue	*Total O_2 consumption (%)*
Adipose	8-10
Brain	18-20
Heart	10-12
Kidney	7-9
Liver	18-22
Muscle	18-22
Other	5-20

Calculated for a 70-kg man at basal metabolism (Wade and Bishop, 1962; Bard, 1961; Baldwin and Smith 1971a).

and should be utilized for balanced appraisal. An early approach (Barcroft, 1947; Bard, 1961) still in extensive use (Reynolds, 1967; Schmidt, 1964) involves determinations of energy expenditures in major tissues and organ systems in basal and other physiological states on the bases of in-vitro and in-vivo rates of oxygen uptake and CO_2 production. A generalized breakdown of energy expenditures in several major tissues is presented in Table 2-2, to provide a perspective for assessing the potential significances of specific physiological processes. For example, energy expenditures associated with an energy-requiring process localized in a single tissue cannot exceed the total energy expenditure of that tissue. Neither can the summed energy expenditures, calculated for a tissue on other bases, exceed the energy expenditure indicated by in-vivo oxygen consumption by that tissue.

Estimates of the percentages of basal energy expenditures accountable for by several specific functions are presented in Table 2-3. Based upon blood flow rates to the kidney and glomerular filtration rates it is estimated that 24-26 moles of Na^+ are reabsorbed per day for a 70-kg man (Guyton, 1961). Of the Na^+ reabsorbed, approximately 65% or 16.25 moles must be reabsorbed by active transport mechanisms. The average efficiency of the renal Na^+ transport mechanisms is on the order of 2.5 Na^+ transported per $\sim$P hydrolyzed (Caldwell, 1966; Kramer and Deetjen, 1964). On this basis, 6.5 moles of $\sim$P would be expended. On

Table 2-3. Energy Expenditures in Several Major Maintenance Functions

Functions	*Basal energy expenditure (%)*
Kidney work (Na transport)	6-7
Protein resynthesis (muscle, liver, intestine)	8-12
Triglyceride resynthesis (adipose, liver)	1-2
Heart work (cardiac output)	9-11
Nervous functions	15-20
Respiration (muscle work)	6-7
Other	30-40

Calculated for a 70-kg man at basal metabolism (1700 Kcal/day); see text for the data.

the average basis of 19 Kcal/~P (justified above), this indicates (6.5 × 19) an expenditure of 123.5 Kcal/day in Na^+ reabsorption in the kidney. This is equal to 7.3% of the daily energy expenditure of a 70-kg man at basal state (Table 2-3). When this estimate is compared with estimates that 7-9% of the daily energy expenditure occurs in kidney (Table 2-2) it can be concluded that a major proportion of the energy expenditure of kidney is associated with Na^+ transport (Kramer and Deetjen, 1964).

Despite the many difficulties inherent in such determinations (Neuberger and Richards, 1964), estimates of rates of turnover of several major proteins and of total body proteins have been made (Neuberger and Richards, 1964; McFarlane, 1964). These are summarized in Table 2-4. Although muscle proteins are quite stable, the mass of muscle protein is very large (4.5-5.0 kg in men); as a result, the amount of muscle protein requiring replacement per day comprises a major proportion of total protein synthesis at maintenance (Table 2-3). The amounts of protein resynthesis per day required for maintenance of muscle, blood, gastrointestinal, and liver proteins account for about three-fourths of the total amount of protein replacement (Table 2-3). Calculating on the bases that 200 g of protein must be replaced per day at maintenance (Table 2-3), that an average amino acid in protein has a molecular weight of 110 g per mole, and that five high-energy pyrophosphate bonds are hydrolyzed in the synthesis of each peptide bond (Baldwin, 1968) releasing (5 × 19.3 Kcal/mole in peptide bond formed) 92 Kcal/mole of peptide bond synthesized, it can be calculated that (200 × 92 ÷ 110) about 167 Kcal of heat are released during resynthesis of proteins broken down each day. This represents about 10% of the basal energy expenditure of a 70-kg man (Table 2-3).

Based upon measurements of the rates of glycerol and free fatty acid release from rat epididymal fat pads incubated in vitro, an assumption that 20% of a rat's body weight is made up of adipose tissue similar to epididymal fat pads, and an estimate that eight ATP pyrophosphate

Table 2-4. Amounts of Resynthesis of Several Proteins and Total Protein in a 70-kg Man (g/day)

Muscle proteins	90-110
Hemoglobin	5-7
Plasma proteins	14-16
Gastrointestinal proteins	15-25
Liver proteins	8-12
Total proteins	150-250

bonds are expended per triglyceride resynthesized (Table 2-5), Ball (1965) calculated that up to 15% of the basal heat production of rats could be attributed to turnover of adipose triglycerides. Ball (1965) recognized the limitation of assuming that all adipose tissue is similar and that in-vitro rates of adipose tissue triglyceride hydrolysis and resynthesis reflect in-vivo rates. Baldwin (1970) applied similar reasoning in tentatively confirming Ball's suggestion that up to 15% of daily energy expenditures could be attributed to triglyceride hydrolysis and resynthesis. The calculations of Baldwin (1970) appear to have been in error since the more recent calculations presented below indicate energy expenditures of only 1.2% of the basal expenditure attributable to triglyceride turnover. According to the metabolic model presented in Table 2-5, (8 × 19) 152 Kcal heat are released per mole of triglyceride hydrolyzed and resynthesized. Determinations in adipose tissue of oxygen uptake associated with triglyceride resynthesis are consistent with this estimate (Ball, 1965). Estimates of amounts of triglyceride resynthesis per day based upon glycerol (0.12-0.13 mg/min/kg) and fatty acid turnover, and oxidation rates determined in resting humans, indicate resynthesis of about 0.14 mole of triglyceride per day (Havel, 1965; Bjorntorp et al., 1969). The indicated energy expenditure is (152 × 0.14) 21 Kcal/day, or 1.2% of basal energy expenditures.

Estimates of energy expenditures associated with heart work, nervous functions, respiration, and other functions presented in Table 2-3 are based upon estimates presented elsewhere (Otis, 1964; Milnor, 1968; Weiner, 1969; Milligan, 1971; Baldwin and Smith, 1971a).

Two general approaches have here been presented to identify energy expenditures associated with maintenance—(1) consideration of energy expenditures by tissues, and (2) consideration of the magnitudes or rates

Table 2-5. Energy Cost of Triglyceride Resynthesis

Model:	Triglyceride + H_2O	→ 3 Fatty acids (FA) + Glycerol	(Adipose)
	Glycerol + ATP	→ ½ Glucose + ADP	(Liver)
	½ Glucose + ATP	→ α-Glycerol-P + ADP	(Adipose)
	3 FA + 3 ATP	→ 3 FA-CoA + AMP	(Adipose)
	3 FA-CoA + α-GP	→ Triglyceride	(Adipose)
Net:	8 ATP (equiv)	→ 8 ADP (equiv)	

Note: Conversion of ATP to ADP involves hydrolysis of one high-energy pyrophosphate bond (~P).

Table 2-6. Energy Expenditures in Liver Functions

Process	*Kcal*	*Liver energy expenditure (%)*
Plasma protein synthesis (18 g/day)	15.7	4.5
Liver protein synthesis (10 g/day)	8.6	2.5
Ribosomal RNA synthesis	0.15	0.045
Na^+/K^+ transport (active)	120.0	35.0
Cori cycle and TG turnover	17.0	5.0
Amino acid degradation*	48.0	14.0
FA conversion to ketone bodies	14.4	4.2
Unaccounted	116.2	34.0

Calculated for liver of a 70-kg man expending 340 Kcal/day in liver function.
*Includes heat loss in glucose, ketone body, and urea synthesis.

of specific energy-requiring processes. In the second approach, it was noted that a major proportion of the energy expended in kidneys is expended in Na^+ transport. Similarly, it has been shown that a major proportion of the energy expenditure of heart is in the performance of work (Milnor, 1968). The liver performs a wide and variable range of functions in support of maintenance. A number of liver functions have been evaluated and estimates of energy expenditures associated with each presented in Table 2-6, in an attempt to further clarify relationships between total tissue energy expenditures and energy expenditures for specific functions within a tissue.

The estimates of costs of plasma protein and liver protein synthesis in liver are based upon estimates of turnover times (Neuberger and Richards, 1964; McFarlane, 1964) of the respective proteins and the same assumptions applied above in calculating energy costs for protein resynthesis in the total animal. The estimate of cost of ribosomal RNA synthesis is based upon a calculation of the energy cost of ribosomal RNA synthesis in rat liver, expressed as a percentage of rat liver energy expenditure (0.045%), which was extrapolated to the human on the assumption that the percentage value calculated for rats applies also to humans (Table 2-6). Assumptions made in the calculation were that the $T_{1/2}$ of rat liver ribosomal RNA is five days (Hirsch and Hiatt, 1966), that liver ribosomal RNA content is 6 mg/g wet weight of tissue, that the average molecular weights of bases in ribosomal RNA is 310 g/mole, and that four high-energy phosphate bonds are hydrolyzed for each base finally appearing in ribosomal RNA. The value obtained is much lower than that preconceived by the authors, a not unusual occurrence. Milligan (1971) calculated that about 35% of the total liver energy expenditure is expended in active ion transport for maintenance of ion concentration gradients, based upon results obtained in vitro with liver slices (Elshove and van Rossum, 1963). The danger resident in quantitative extrapolation of in-vitro tissue-slice data is recognized, but the arguments presented by Milligan (1971) and others (Whittam, 1964; Schotellius and Schotellius, 1968; Keynes and Maisel, 1954; Levinson and Hempling, 1967; Caldwell, 1968), suggesting that energy expenditures in active transport of Na^+/K^+ are large in accounting for about 50% of brain energy expenditures in vitro, 10-20% of resting muscle energy expenditures, and 85% of the energy expenditures of ascites tumor cells, are good and of sufficient validity to support the contention that significant amounts of energy are expended in the active transport of ions in many tissues. The estimates of liver energy expenditure in glucose resynthesis from lactate (Cori cycle) and glycerol (triglyceride turnover) in Table 2-6 are based upon a previous formulation (Baldwin, 1968), and the estimate of triglyceride turnover rates presented above. The estimate of costs of amino acid degradation (Table 2-6), is based upon amino acid requirements for maintenance and calculations of heat losses during the conversion of mixed amino acids to glucose, ketone bodies, urea, and other degradation products. The assumptions used in the calculations of these heat losses were those formulated by Krebs (1964) and discussed by Baldwin and Smith (1971a). The estimated heat loss in the conversion of fatty acids removed from blood by liver to ketone bodies was based upon estimates of ketone body production rates in humans (McPherson et al., 1958) and other species (Katz and Bergman, 1969).

The specific energy expenditures presented in Table 2-6 leave 34% of the energy expenditures of liver unaccounted for. With the exceptions of tissues such as heart, kidney, and mammary, where major proportions of tissue energy expenditures are accountable for on the basis of specific major functions such as work, ion transport, or milk synthesis, most attempts to account for specific energy expenditures in tissues are about 60-70% successful, as indicated for liver in Table 2-6. The 30-40% unaccounted energy expenditures of tissues may (1) represent a failure to identify several significant functions in each tissue, (2) indicate that current data do not provide for identification of several significant functions of each tissue, (3) be a reflection of the possibility that 30-40% of the energy expenditures in many tissues are attributable to a large number of functions such as ribosomal RNA synthesis which are difficult to account for, or (4) indicate all three of these possibilities. Further study should lead to identification of all of the molecular processes which determine the energy required for maintenance of the animal.

Heat Increments

Heat Increments of Feeds at Maintenance

Formulations directed at identifying and evaluating the processes of digestion and assimilation which contribute to the heat increment of maintenance (HI_m) have been discussed elsewhere (Krebs, 1964; Baldwin, 1968, 1970; Baldwin and Smith, 1971a), and hence will not be discussed in detail here. Estimates of the heat increments of protein meals were based upon the pathways of amino acid degradation and of conversion of intermediates to ketone bodies, glucose, and excretory products, including urea (Krebs, 1964; Baldwin and Smith, 1971a). These estimates account for heat increments of 20-27%, and agree well with experimental results (Brody, 1945). Estimates of the contributions of several factors to the heat increments of carbohydrate and carbohydrate plus fat meals are summarized in Table 2-7 (Baldwin, 1968, 1970).

The major factors contributing to the HI_m of carbohydrate meals appear to be absorption (2.6%) and assimilation (5.7%). The primary costs of assimilation are due to energy expended in glycogen and fat synthesis from glucose for the purpose of storage. Lesser energy losses are associated with replacement of digestive proteins secreted during digestion, bond breakage during digestion, and other processes not listed in Table 2-7, such as increased cardiac work. The percentage of energy fed as triglyceride expended in absorption and assimilation (incorporation into body triglyceride stores) is quite small (Table 2-7; Baldwin, 1970). The estimates presented in Table 2-7 agree well with direct estimates (Brody, 1945).

Heat Increments of Production

Since net energy values of feedstuffs provided in excess of maintenance requirements for production are, by definition, equivalent to the amounts of energy gained in the animal's body or secreted in milk (Armsby, 1917), heat increments of production can be considered collective terms which might be expected to include increased costs for digestion and assimilation, increased costs for a variety of physiological functions associated with the productive process, and costs of synthesis of components of gain (protein, fat, etc.) and milk (protein, fat, lactose). In young, normal animals, net energy values of well-balanced diets provided for growth and for lactation range between 55-65% and 60-73%, respectively, of the metabolizable energy (ME) contents of the diets (Blaxter, 1962; Flatt et al., 1969). This means that heat increments for growth range between 35 and 55%, and for lactation, between 25 and 40%.

The same formulations, applied in estimating heat increments of diets fed for maintenance (Baldwin, 1968; Table 2-7), can be applied in estimating energy losses associated with digestion and assimilation of ME

Table 2-7. Estimated Heat Increments of Two Meals Fed at Maintenance Level and of Corresponding Heat Losses Associated with the Utilization of Metabolizable Energy Provided for Production Above Maintenance Level (%)

Process	*Carbo-hydrate meal*	*Mixed meal*	*Above maintenance level*
Bond breakage	0.5	0.3	0.3-0.4
Absorption (active transport)	2.6	1.3	1-2.5
Digestive secretions*	2.0	2.0	1-2
Assimilation and/or storage†			1-2
Glucose as glycogen (0.3 × 5%)	1.5	1.5	
Glucose as fat (0.5 × 8.2%)	4.2	—	
Fat as fat (0.5 × 3.0%)	—	1.5	
Theoretical HI_m	10.8	6.6	3.3-7.0
Observed HI_m	9-12	5-7	25-45

Values are expressed as percentage of metabolizable energy fed as carbohydrate, carbohydrate plus fat (50:50), or an increment of balanced ration fed for production above maintenance level and lost as heat during digestion and assimilation.

*Includes cost of synthesis of the portion of digestive proteins replenished in the period after feeding during which heat increment is evident.

†Assumes, in the case of diets fed at maintenance levels, that 80% of calories provided in the diet are stored as glycogen (30%) or triglyceride (50%). Estimated energy expenditures for conversions of glucose to glycogen, glucose to storage triglyceride, and dietary triglyceride to storage triglyceride are 5.0%, 8.2%, and 3.0%, respectively (Baldwin, 1968).

provided for production above maintenance levels. For a mixed, balanced diet, energy losses due to bond breakage, costs of active transport in the absorption of nutrients, synthesis of digestive secretions, and assimilation or storage of nutrients would be expected to range between 0.3 and 0.4%, 1 and 25%, 1 and 2%, and 1 and 2%, respectively, of ME (Table 2-7). The decreased estimate for assimilation or storage costs for production, as compared to maintenance, (Table 2-7) is because productive animals tend to eat more frequently, and a greater proportion of absorbed nutrients is utilized directly for productive processes. The formulation above indicates that only between 3.3 and 7.0% of the ME provided for production is expended in digestive and assimilative processes, and, therefore, that these processes account for only a small proportion of heat increments of production (Table 2-7).

It is difficult to account for the increased costs by the variety of physiological functions that might be elevated in producing animals. Considerations of increased costs attributable to increased endocrine activities, increased cardiac output, increased activity, etc., suggest that the identifiable costs of increased physiological activities comprise only a small proportion of heat increments of production. A notable exception to this generalization is the increased maintenance requirement observed in lactating animals by Brody (1945) and substantiated further by Flatt et al. (1969). This increased maintenance requirement appears to reflect a general increase in basal metabolic activity, and may be due to increased secretion of thyroxin, glucocorticoids, prolactin, and growth hormone in lactating animals (Cowie, 1969; Baldwin and Smith, 1971a,b). However, this increase in heat production, associated with general physiological changes during lactation, has been considered a maintenance function by workers in nutritional energetics (Brody, 1945; Flatt et al., 1969), and is not, therefore, part of heat increment estimates for lactation. A major portion of the heat increments of growth, fattening, and lactation can be accounted for as costs of biosynthesis of the carbohydrates, proteins, and lipids appearing in milk and as body gain (Baldwin, 1968). Summary formulations estimating energy costs of lactose, protein, and triglyceride synthesis are presented in Table 2-8.

The formulations are based upon known metabolic and biosynthetic pathways (Mahler and Cordes, 1966; Baldwin, 1968). Estimates are expressed on the basis of efficiency of incorporation of metabolizable energy into products as:

$$\frac{\text{Energy in product}}{\text{ME input}} \times 100 = \frac{NE_P}{ME} \times 100 = \%\ \text{efficiency}$$

Table 2-8. Summary Formulations of Energy Costs of Lactose, Protein, and Triglyceride Synthesis from Products of Ruminant Digestion

---

Lactose synthesis from propionate

Conversion of propionate to glucose in liver
4 propionate + 8 ~ P → 2 glucose

Transport of glucose to and synthesis of lactose in udder
2 glucose + 4 ~ P → lactose

Calculation of efficiency
output: 1 lactose = 1,350 Kcal/mole
input: 4 propionate @ 367 Kcal/mole + 12 ~ P @ 21 Kcal

$$\text{Efficiency} = \frac{1350}{1720} \times 100 = 78\%$$

Protein synthesis from balanced amino acids

Balance equation
1.0 amino acids + 5 ~ P → 100 g protein

Calculation of efficiency
Output: 100 g protein @ 570 Kcal/100 g
Input: 1.0 mole mixed amino acid @ 624 Kcal/mole + 5 ~P @ 21 Kcal

$$\text{Efficiency} = \frac{627}{729} \times 100 = 86\%$$

Triglyceride synthesis from acetate*

Palmitate synthesis
8 acetate + 23 ~ P + 14 $NADPH_2$ → Palmityl-CoA + 14 NADP

Formation of $NADPH_2$ from propionate via glucose and the pentose phosphate pathway
2.34 propionate + 4.68 ~ P → 1.17 glucose (liver)
1.17 glucose + 2.34 ~ P → 1.17 G6P (mammary)
1.17 G6P + 14 NADP → 14 $NADPH_2$ (mammary)

Triglyceride formation:
propionate + 6 ~ P (equiv.) → α-glycerol-P
α-glycerol-P + 3 palmityl CoA → triglyceride

Calculation of efficiency:
Output: tripalmitate @ 7,600 Kcal/mole
Input: 24 acetate @ 209 Kcal/mole + 8.02 propionate @ 367 Kcal/mole + 96 ~ P @ 21 Kcal/mole

$$\text{Efficiency} = \frac{7{,}600}{9{,}975} \times 100 = 76\%$$

---

*Since acetate is a major energy source in ruminants, 21 Kcal/mole of ~ P was utilized as for the equivalent of ~ P in these calculations (see Table 2-1).

Table 2-9. Estimation of Efficiency of Synthesis of 100 g of Milk

Milk component	*%*	*Kcal output*	*Estimated efficiency**	*Estimated ME required above maintenance*
Protein	3.3	19.0	0.86	22.1
Fat	3.7	34.6	0.76	45.5
Lactose	4.8	19.2	0.78	24.6
Net		72.8	0.79	92.2

*Based on values in Table 2-8.

The primary precursors of milk lactose, protein, fatty acids, and triglyceride glycerol were considered to be absorbed propionate, amino acids, acetate, and propionate, respectively. The actual comparison of input to metabolizable energy in these equations does not apply to the determined ME value of a feedstuff unless metabolizable energy values determined for that feedstuff are corrected for heat released during rumen fermentation (Blaxter, 1962). The calculated efficiency values of 78%, 86%, and 76%, respectively, for lactose, protein, and triglyceride synthesis must be considered maximal estimates since only direct costs of synthesis are considered. Nevertheless, when these values are applied in estimating efficiencies of utilization of corrected ME for lactation and growth, values of 79% and 81%, respectively, are obtained (Tables 2-9 and 2-10). These estimates indicate, when compared to minimum reported heat increments of lactation and growth of 27% and 35% (Blaxter, 1962, and Flatt et al., 1969, respectively), that (21 × 100/27) 78% and (19 × 100/35) 54% of the heat increments of lactation and growth, respectively, can be accounted for by considering only direct energy expenditures required in the biosynthesis of the major components of milk and gain. It does not seem overly optimistic to suggest that consideration of additional functions such as increased cardiac work, additional molecular transformations requiring energy, etc., will result in an accounting of the presently unaccounted for portions of the heat increments of growth and lactation. In a detailed computer-assisted simulation analysis of milk synthesis in the mammary gland, it was found that more than 90% of the energy expenditures of mammary glands in milk synthesis could be accounted in specific functions (Baldwin and Garfinkel, unpublished; Barry, 1964; Linzel, 1967; Reynolds, 1967).

Table 2-10. Estimation of Efficiency of Growth in Young Lambs

	Composition of gain (g)	*Gain (Kcal)*	*Estimated efficiency**	*Estimated ME required above maintenance*
Protein	926	5,278	0.86	6,137
Fat	414	3,880	0.76	5,105
Total	1,340	9,158	0.81	11,242

*Based on values in Table 2-8.

As discussed by Blaxter (1962), Baldwin (1968), and Milligan, (1971), the major challenge facing investigators attempting to relate aspects of biochemistry, metabolism, and physiology to the nutritional energetic terms defined so long ago (Armsby, 1917) is an explanation of the higher heat increments of 35 and 45% in average lactating and growing animals, and the extreme heat increments for fattening of 60-70% sometimes observed in adult animals (Blaxter, 1962). Identification of the factors contributing to these reduced efficiencies might well provide bases for improving efficiencies of feedstuff utilization for production.

References

Annison, E. F., Brown, R. E., Leng, R. A., Lindsay, D. B., and West, C. E., 1967. Rates of entry and oxidation of acetate, glucose, D(-)-ß-hydroxybutyrate, palmitate, oleate and stearate, and rates of production and oxidation of propionate and butyrate in fed and starved sheep. *Biochem. J.* 104:135.

Armsby, H. P., 1917. *The Nutrition of Farm Animals.* The Macmillan Company, New York.

Baldwin, R. L., 1968. Estimation of theoretical calorific relationships as a teaching technique. A review. *J. Dairy Sci.* 51:104.

Baldwin, R. L., 1970. Metabolic functions affecting the contribution of adipose tissue to total energy expenditure. *Fed. Proc.* 29:1277.

Baldwin, R. L., and Smith, N. E., 1971a. Application of a simulation modeling technique in the analysis of dynamic aspects of animal energetics. *Fed. Proc.* 30:1459.

Baldwin, R. L., and Smith, N. E., 1971b. Intermediary aspects and tissue interactions of ruminant fat metabolism. *J. Dairy Sci.* 54:583.

Ball, E. G., 1965. Some energy relationships in adipose tissue. *Ann. N. Y. Acad. Sci.* 131:225.

Barcroft, S. J., 1947. *Researches on Pre-natal Life*. Charles C Thomas, Springfield, Ill.

Bard, P., 1961. O_2 uptake and tissue energy expenditure. In *Medical Physiology* (P. Bard, ed.). C. V. Mosby Company, St. Louis, p. 239.

Barry, J. M., 1964. A quantitative balance between substrates and metabolic products of the mammary gland. *Bio. Rev.* 39:194.

Bjoorntorp, P., Bergman, H., Varnauskas, E., and Lindholm, B., 1969. Lipid metabolism in relation to body composition in man. *Metabolism* 18:840.

Black, A. W., 1970. Nitrogen and carbohydrate metabolism discussion, in *Physiology of Digestion and Metabolism in the Ruminant* (A. T. Phillipson, ed.). Oriel Press, Cambridge, p. 452.

Blaxter, K. L., 1962. *The Energy Metabolism of Ruminants*. Charles C Thomas, Springfield, Ill.

Brody, S., 1945. *Bioenergetics and Growth*. Reinhold, New York.

Caldwell, P. C., 1966. Energy relationships and the active transport of ions. In *Current Topics in Bioenergetics* (D. R. Sanadi, ed.). Academic Press, New York, p. 251

Caldwell, P. C., 1968. Movement and distribution of inorganic ions. *Physiol. Rev.* 48(1):13.

Cowie, A. J., 1969. General hormonal factors involved in lactogenesis. In *Lactogenesis: The Initiation of Milk Secretion at Purification* (M. Reynolds and S. J. Folley, eds.). University of Pennsylvania Press, Philadelphia, p. 157.

Defares, J. G., 1964. Principles of feedback control and their application to the respiratory control system. In *Handbook of Physiology, Sec. 3—Respiration, Vol. I* (W. O. Fenn and H. Rohn, eds.). Am. Physiol. Soc., Washington, p. 649.

Elshove, A., and van Rossum, G. V. D., 1963. Net movements of sodium and potassium and their relation to respiration, in slices of rat liver incubated in vitro. *J. Physiol.* 168:531.

Flatt, W. P., Moe, P. W., Munson, A. W., and Cooper, T., 1969. Energy utilization by high producing dairy cows. II. Summary of energy balance experiments with lactating Holstein cows. In *Energy Metabolism of Farm Animals* (K. L. Blaxter, J. Kielanowski, and G. Thorbeck, eds.). OrieAriel Press, New Castle, England, p. 235.

Garfinkel, D., Frenkel, R. A., and Garfinkel, L., 1968. Simulation of the detailed regulation of glycolysis in a heart supernatant preparation. *Computers and Biomedical Res.* 2:68.

Guyton, A. C., 1961. Formation of urine by the kidney. In *Textbook of Medical Physiology*. W. B. Saunders Company, Philadelphia, p. 83.

Havel, R. J., 1965. Some influences of the sympathetic nervous system and insulin on mobilization of fat from adipose tissue: Studies of the turnover rates of free fatty acids and glycerol. *Ann. N. Y. Acad. Sci.* 131:91.

Hirsch, C. A., and Hiatt, H. H., 1966. Turnover of liver ribosomes in fed and in fasted rats. *J. Biol. Chem.* 241:5936.

Jobsis, F. F., 1964. Basic processes in cellular respiration. In *Handbook of Physiology, Sec. 3—Respiration, Vol. I* (W. O. Fenn and H. Rohn, eds.). Am. Physiol. Soc., Washington, p. 63.

Katz, M. L., and Bergman, E. N., 1969. Hepatic and portal metabolism of glucose, free-fatty acids, and ketone bodies in the sheep. *Am. J. Physiol.* 216:953.

Keynes, R. D., and Maisel, G. W., 1954. The energy requirement for sodium extrusion from a frog muscle. *Proc. Roy. Soc.* B142:383.

Kramer, K., and Deetjen, P., 1964. Oxygen consumption and sodium reabsorption in the mammalian kidney. In *Oxygen in the Animal Organism.* I.U.B. Symposium Series, Vol. 31 (F. Dickens and E. Neil, eds.). The Macmillan Company, New York, p. 411.

Krebs, H. A., and Kornberg, H. L., 1957. Energy transformations in living matter. *Ergeb. Physiol.* 49:212.

Krebs, H. A., 1964. The metabolic fate of amino acids. In *Mammalian Protein Metabolism* (H. N. Munro and J. B. Allison, eds.). Academic Press, New York, p. 125.

Levin, R. J., 1969. The effects of hormones on the absorptive, metabolic and digestive functions of the small intestine. *J. Endocr.* 45:315.

Levinson, C., and Hempling, H. G., 1967. The role of ion transport in the regulation of respiration in the Ehrlich mouse ascites-tumor cell. *Biochim. Biophys. Acta* 135:306.

Linzell, J. L., 1967. The magnitude and mechanisms of the uptake of milk precursors by the mammary gland. *Proceedings of the National Institute for Research in Dairying.* 27:44.

Mahler, H. R., and Cordes, E. H., 1966. *Biological Chemistry.* Harper and Row, New York.

McFarlane, A. S., 1964. Metabolism of plasma proteins. In *Mammalian Protein Metabolism* (H. N. Munro and J. B. Allison, eds.). Academic Press, New York, p. 298.

McPherson, H. T., Werk, E. E., Jr., Myers, J. D., and Engle, F., 1958. Studies on ketone metabolism in man. II. The effect of glucose, insulin, cortisone and hypoglycemia on splanchnic ketone production. *J. Clin. Invest.* 37:1379.

Milligan, L. P., 1971. Energy efficiency and metabolic transformations. *Fed. Proc.* 30:1454.

Milnor, W. R., 1968. Blood supply of special regions. In *Medical Physiology* Vol. I (V. B. Mountcastle, ed.). C. V. Mosby Company, St. Louis, p. 221.

Neuberger, A., and Richards, F. F., 1964. Studies on turnover in the whole animal. In *Mammalian Protein Metabolism* (H. N. Munro and J. B. Allison, eds.). Academic Press, New York, p. 243.

Otis, A. B., 1964. Quantitative relationships in steady-state gas exchange. In *Handbook of Physiology, Sec. 3—Respiration, Vol. I.* (W. O. Fenn and H. Rohn, eds.). Am. Physiol. Soc., Washington, p. 681.

Reynolds, M., 1967. Mammary respiration in lactating goats. *Am. J. Physiol.* 212:707.

Schmidt, C. F., 1964. Cerebral blood supply and cerebral oxidative metabolism. In *Oxygen in the Animal Organism.* I.U.B. Symposium Series, Vol. 31 (F. Dickens and E. Neil, eds.). The Macmillan Company, New York, p. 443.

Schottelius, B. A., and Schottelius, D. D., 1968. Basal energy utilization in mammalian skeletal muscle. *Proc. Soc. Exptl. Biol. and Med.* 127:1228.

Wade, O. L., and Bishop, J. M., 1962. *Cardiac Output and Regional Blood Flow.* Blackwell, Oxford.

Weiner, D. E., 1969. Cerebral blood flow measurement by external isotope detection of ^{131}Cs. *J. Appl. Physiol.* 27:556.

Whittam, R., 1964. *The Cellular Functions of Membrane Transport* (J. F. Hoffman, ed.). Prentice-Hall Inc., Englewood Cliffs, N. J., p. 139.

3

Control of Carbohydrate and Fat Metabolism

Jay Tepperman

Since the public and the editors of the mass media have become obsessed with the idea of ecology it is gratifying to realize that we metabolic physiologists have been ecologists all along. Anatomically, the lumen of the gastrointestinal tract is outside the body. Ultimately, the controlling signals for carbohydrate, fat, and protein metabolism, as well as for water and electrolyte balance, either originate or fail to originate in the gastrointestinal tract. Surely, there is no more impressive demonstration of the environment-organism continuum than the one that can be contemplated from the vantage point of the brush border of an intestinal epithelial cell. Respiratory physiologists, who regard the alveolar membrane as their boundary of self, can see things happen faster than we can, but adaptive responses to starvation and refeeding, and accommodations to variations in dietary composition, are no less impressively coordinated than are responses to hypoxia.

At all levels of biological organization the most striking feature of metabolic control mechanisms is their remarkable symmetry. Depending on one's preference for German philosophy, Greek drama, or ancient oriental wisdom, metabolic controls can be described in terms of thesis and antithesis, agonist and antagonist, or Yang and Yin. Almost always, one finds a balance between opposing forces in the steady-state circumstance. If the steady state is perturbed in either direction, one or the other set of control mechanisms is brought into play and gains ascendancy. Often, the response is a coordinate one; i.e., one set of controls is activated while the opposing set is inhibited. However, I propose to show at least one example in which both agonist and antagonist are simultaneously stimulated to achieve a notably sensible overall effect.

The control processes that are set in play by the ingestion of food can be regarded as a cascade of sequential signals. I hope to demonstrate that, at least in the case of selected individuals, some of these signals arise in the central nervous system even in the absence of food in the gastrointestinal tract. When carbohydrates, fats, protein hydrolysates, or (in the case of ruminants) volatile fatty acids are absorbed, these substances themselves constitute a primary set of signals. Depending on the composition of the metabolic mixtures, the primary signals elicit a mix of hormonal signals, some arising from the gastrointestinal tract and others from the pancreatic alpha and beta cells. The intricate interplay of neural and hormonal signals, and the primary signals in the form of nutrients absorbed from the intestine, then determine the substrate composition of the circulating blood. It should be emphasized that substrate concentrations constitute information, and, therefore, participate in metabolic control, just as do the partial pressure of CO_2 and O_2 in the control of respiration. Insulin and glucagon, working in collaboration with other permissive hormones and the metabolic mixture, then react with their cellular recognition sites and initiate appropriate propagated disturbances which involve highly coordinated adjustments of substrate flow patterns. Since some intracellular control mechanisms are discussed in other chapters of this volume, and since we have already summarized some of our reflections on this subject (Tepperman and Tepperman, 1970), I propose to stay out of the cell on this occasion.

In order to provide models for some of the ideas I have been sketching, I plan to construct a montage of data, most of it gathered by other people—some of it quite recently. I shall deal with only two general topics: (1) the role of the central nervous system in the release of insulin and (2) the insulin-glucagon pair and the regulation of their secretion by nonneural signals.

The Central Nervous System and Insulin Release

I was brought up to believe that the major influence of the central nervous system on metabolism was by way of its participation in the regulation of food intake. In retrospect, evidence of a more direct metabolic controlling influence can be found in the literature as far back as Claude Bernard's piquant diabetes experiment, in studies at the turn of the century, best summarized in John Brobeck's Ph.D. thesis (1939), in Britton's 1925 report, and in a number of clinical studies in which diabetes was made manifestly more severe by psychological trauma (Hinkle et al., 1951). However, it was not until the availability of a radioimmunoassay for insulin that earlier suspicions of direct participation by the central nervous system (CNS) in the release of insulin were confirmed.

One of the most striking demonstrations of CNS modulation of insulin secretion can be found in the experiments of John Mason and his colleagues on the effects of psychological trauma in the monkey on many hormonal responses (Mason, 1968). These investigators regularly found a small but significant fall in circulating immunoreactive insulin in association with avoidance conditioning and other types of psychological stress. Sometimes the fall in insulin levels during stress was followed by a compensatory overshoot during the recovery period. This series of experiments reminds us once again of the highly coordinated nature of the stress response: certain hormones are elicited while others are suppressed. It also serves to underscore the fact that there is a huge body of information concerned with the metabolic effects of glucocorticoids, catecholamines, thyroid hormone, growth hormone, and sex steroids, which we are simply ignoring in this discussion.

Since (as we shall see again) catecholamines inhibit insulin release, Mason and his coworkers wondered whether the mechanism of inhibition of insulin secretion associated with stress could be the often-noted increase in catecholamine hormone production. Since they saw decreases in circulating insulin levels in animals which did not show changes in plasma epinephrine concentration, they rejected this hypothesis.

It is interesting to contemplate an experiment of Han et al. (1970) in the light of Mason's findings. These investigators strongly suspected that ventromedial hypothalamic nuclear lesions not only cause hyperphagia but may also produce a relative hyperinsulinism independent of food intake. Therefore, they examined the Islets of Langerhans in two groups of hypophysectomized rats, one serving as controls and one with ventromedial lesions. Both groups of rats were fed rigidly controlled and equal amounts of food by stomach tube. They found a significant increase in mean islet area in the lesioned group of rats but, perhaps more strikingly, they found that the calculated volume of the two largest islets in each animal was significantly higher in the lesioned group. If the CNS exerts tonic control over the pancreatic beta cell, it is extraordinarily interesting that a part of the brain demonstrably implicated in food intake regulation may also be involved in beta cell control.

Begging the question of the nature of the brain circuitry involved in modulation of insulin release, how does the information generated in the brain reach the beta cell? It is now generally agreed (Frohman et al., 1966; Kaneto et al., 1967) that stimulation of the vagus produces insulin release. My favorite experiment, which illustrates this point, is that of Daniel and Henderson (1967), who stimulated the distal cut ends of the vagi in baboons and monitored immunoreactive insulin in the serum. Clearly, this procedure produces a rise in circulating insulin. Whether this is in part a vascular response is not known, though one can release insulin from pancreas slices with carbaminoylcholine, as Coore and Randle (1964) have shown.

Aesthetically it would be pleasing if epinephrine and norepinephrine had an effect on insulin release opposite to that of acetylcholine. Porte and Williams (1969) have demonstrated such to be the case in intact man. It has also been demonstrated in a number of in-vitro systems as well (Malaisse et al., 1967). Since the beta cell is now believed to be directly innervated by both vagal and sympathetic nerves (Esterhuizen et al., 1968), the net neural effect on insulin secretion would depend on the balance of the two inputs.

Some time ago, in collaboration with Walter Allan, I became interested in the possibility that insulin secretion might be conditionable by a taste stimulus. While we were doing our first experiments Goldfine et al. (1969) published their findings on a group of four normal subjects given glucose-cola ("Glucola"), diet-cola, and water on three different occasions. They concluded that artificially sweetened cola did not cause an increase in immunoassayable insulin.

We studied only two subjects; the first, a 25-year-old man, completely confirmed the conclusions of the Goldfine group. We had considered Glucola but rejected it on the grounds that it contained caffeine, which might itself have some influence on insulin release. Our first subject drank lemon juice, with glucose on one occasion, and with artificial sweetener (cyclamate-saccharin, "Sucaryl") on another. The subject was unable to distinguish one preparation from the other by taste or appearance. Glucose produced a large increase in immunoassayable insulin, and Sucaryl resulted in no change for two hours.

The second subject was a 54-year-old man who followed the same procedure (Figure 3-1). Again hyperglycemia associated with glucose ingestion was associated with a large and sustained increase in circulating insulin concentration. Sucaryl in lemon juice, also taken blind, caused a less striking rise, while 300 ml of water produced no change.

At this point we wondered whether the Sucaryl itself, apart from its gustatory effect, would cause an increase in immunoreactive insulin, and whether lemon juice might do so as well, possibly by causing release of some gastrointestinal hormone. Therefore, the subject ingested 350 mg of Sucaryl in a gelatin capsule on one occasion and 300 ml of unsweetened lemon juice on another. Neither of these maneuvers caused a change in resting insulin concentration (Figure 3-1).

The final experiment of the summer was somewhat marred by the fact that the subject knew he was drinking the artificially sweetened lemonade. A summary of all experiments done on this subject is shown in Figure 3-2.

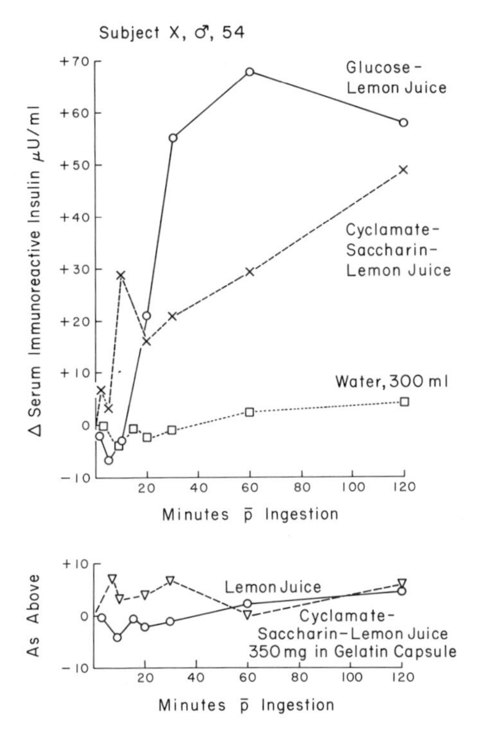

Figure 3-1. Effect of taste stimulus on changes in serum immunoreactive insulin in a 54-year-old subject (see text for discussion).

With only a few exceptions, all of the values of the Sucaryl test days fell outside the range of values observed on all the control days.

Understandably, we are diffident about drawing sweeping conclusions from our series of one case. However, our experience has stimulated us to ask a number of questions which may be of more than passing interest to clinicians who are interested in functional hypoglycemia. If the response shown by our second subject is idiosyncratic, how frequently does it occur in a large sample of the population? Could this type of reaction be a component of clinical hyperfunction of the beta cell in some patients? If only one of Goldfine's four subjects was a responder to the taste stimulus, could his response have been masked by pooling all four subjects in one group?

Shortly after these events, Goldfine and his colleagues provided us with our most entertaining proof of the participation of the CNS in insulin release (Goldfine et al., 1970). This team of investigators hypnotized

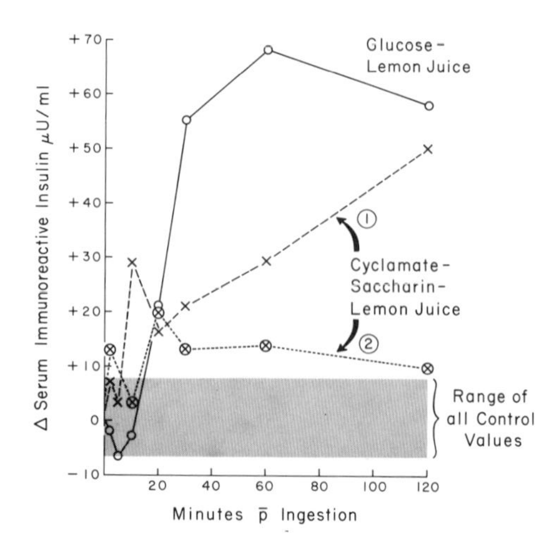

Figure 3-2. Summary of studies on the effect of taste stimulation on changes in serum immunoreactive insulin; same subject as in Fig. 3-1.

seven individuals and, while they were in a state of hypnosis, suggested to each that he was in his favorite restaurant enjoying his favorite meal. In three of them, a substantial increase in serum immunoreactive insulin was observed in response to hypnotic suggestion! It is noteworthy that the increased circulating levels of insulin were not accompanied by hypoglycemia. A possible mechanism for this response will be suggested later.

Substrate-Induced Insulin Release

Until a few years ago, many of us confidently taught that glucose is the stimulus for insulin release. This was consistent with the assumption that insulin had something to do with carbohydrate metabolism. Historically, the association of insulin with carbohydrate metabolism was due largely to the fact that diabetics are (by definition) glycosurics. The panmetabolic nature of insulin's action and the complexity of the metabolic repercussions of its lack were well established, but the signal remained simple and it remained the same. When the problem was studied with radioimmunoassay methods, the old impression that glucose is a powerful signal for release of insulin was nicely confirmed. For example, in the isolated, perfused rat pancreas, glucose provokes insulin release in proportion to the signal strength—i.e., glucose concentration (Sussman et al., 1966). The relationship to signal strength can be appreciated even better in such experiments as Coore's and Randle's (1964) in which the insulin output by pancreas slices of the rabbit is related to the glucose concentration of the medium.

It is difficult to sort out the historical developments which led to our present knowledge of the variety and complexity of insulin release signals. Certainly the reintroduction of the oral antidiabetic drugs stimulated interest in the concept of insulin release. The seminal discovery by Fajans and his colleagues (Floyd et al., 1966) that certain amino acids and protein hydrolysates can increase serum immunoreactive insulin (IRI) (we will see an example later) was another landmark in the analysis of insulin release. For me, the arresting demonstration by McIntyre et al. (1965) that oral glucose provokes a greater increase in serum IRI than does intravenous glucose (though the latter results in a much higher level of circulating glucose concentration) clearly indicated that blood glucose level cannot be the only stimulus to insulin release. The demonstration that volatile fatty acids in ruminants are better stimuli of insulin release than is glucose (Manns et al., 1967; Horino et al., 1968) made considerable teleologic sense since the circulating glucose levels in ruminant species do not fluctuate very widely. Although Stern et al. (1970) have recently questioned the significance of volatile fatty acids as beta cell signals, I confess that I shall be saddened if they prove to be right. In addition, fatty acids and keto acids have been shown to have insulin-releasing ability (Williams and Ensinck, 1966), though it is difficult at present to understand the physiologic role of such signals in the total economy of the animal.

Other Hormones as Insulin-Release Stimuli

From an analysis of all of the substrate signals for insulin release, there has evolved a well-worn concept of insulin as an all-purpose anabolic hormone, whose main role is not simply to facilitate glucose transport but to promote storage of nutrients as glycogen, fat, or protein during meals. It is interesting, then, that the McIntyre experiment suggests the existence of signals related to the act of eating. Possibly these signals could contain neural components which involve informational afferents to the CNS, information processing, and the sending of appropriate, coordinated efferent messages to effect insulin release.

The discovery that glucagon stimulates the release of insulin (Samols et al., 1965), and that this effect was magnified by theophylline, led to the demonstration that 3′, 5′-cyclic AMP is the probable agent by which insulin release is brought about by many peptides (Mayhew et al., 1969). Prominent among these are the gastrointestinal hormones secretin and pancreozymin, which share some of glucagon's primary structure. An experiment reported by Allan and Tepperman (1969) illustrates elevation of serum IRI by pancreozymin and aminophylline in the rat and a striking increase when both were given together. This suggests that cyclic 3′, 5′-adenosine monophosphate (AMP) is involved in the response.

The complexity of the forces at play on the beta cell during the absorption of food from the gastrointestinal tract is well illustrated by the studies of Kraegen et al. (1970) on the modification of the insulin response to intravenous glucose infusion, by a prior infusion of secretin. First they demonstrated the consistency of the glucose and insulin responses in two subjects on different days. Then they infused these two subjects with secretin 7 or 25 minutes before the second glucose infusion. It is apparent that the secretin produced a small but perceptible increase in serum IRI even before the second glucose infusion. Even more striking is the exaggerated and sustained insulin response in the secretin-pretreated subjects. Finally they demonstrated, by specific radioimmunoassay, that the secretin itself disappeared very quickly from the circulation. They are left with the hypothesis that oral glucose itself functions as an insulin release signal but that, at the same time (or even before, via other, potentiating effects), it provokes the secretion of secretin. The secretin transitorily exerts a direct stimulatory effect on the beta cell, but, more importantly, it somehow enhances the beta cell's response to the glucose stimulus.

The Control of Glucagon Secretion

Until recently, people who worked on pancreatic glucagon were often apologetic because they were not sure that it has a physiologic role. Even after the study of fluctuations in circulating glucagon levels by radioimmunoassay was introduced, there were many discrepant findings by different groups of investigators, possibly because of the cross-reaction between antiglucagon antibody and gastrointestinal glucagon. Glucagon turned out to be the glamor hormone of 1970, for there is now a clearly demonstrable glucagon deficiency state and hyperglucagonemia is now believed to be a common feature of uncontrolled diabetes mellitus in man. Much of the reason for these advances has been due to the development of an immunospecific antibody for pancreatic glucagon by Unger and his colleagues (1967).

An unequivocal demonstration of a glucagon deficiency state was described by Grey et al. (1970). In a beautifully simple experiment these investigators showed that 48-hour-fasted rats injected with glucagon antiserum develop a substantial hypoglycemia (a 45 mg/100 ml fall in blood glucose) within 75 minutes. This certainly suggests strongly that the circulating glucagon in a 48-hour-fasted rat is necessary for sustaining gluconeogenesis at a sufficiently high level to maintain the blood glucose. There had been older, inferential evidence to suggest that glucagon is essential for adaptation to starvation, but nothing as attractive as this.

The coordinate control of the insulin-glucagon mix has been well studied by Unger and his colleagues. A summary of their interesting work describes the insulin and glucagon responses to predominantly carbohydrate or protein meals (Muller et al., 1970). In the case of normal subjects, a carbohydrate meal results in hyperglycemia and hyperinsulinemia. When these changes are well established, the circulating levels of pancreatic glucagon decrease significantly. Note that the glucagon concentrations are reported in $\mu\mu$ g/ml!

In contrast to the normal response, a group of diabetics showed a marked impairment of glucose tolerance, an obtunded insulin response, and no suppression of the circulating glucagon levels. Unger points out that nonsuppressibility of the pancreatic alpha cell may be one of the central features of the diabetic state and can clearly result in overproduction of glucose from amino acids, even while carbohydrate is being absorbed from the gastrointestinal tract. Since glucagon is immediately suppressible in diabetic dogs given insulin (Muller et al., 1970), Unger's group has suggested that the alpha cell may be insulin-sensitive. There is no doubt that this intriguing suggestion will be subjected to further experimental testing.

When Fajans and his colleagues first described a striking elevation in circulating insulin levels in response to the ingestion of a meat meal (Floyd et al., 1966), it was difficult to understand why the hyperinsulinemia did not result in hypoglycemia. Now, Unger and his colleagues (Muller et al., 1970) have provided us with the explanation. For a meat meal, in addition to serving as a powerful signal for insulin release, also stimulates the release of glucagon. Thus, the blood glucose-lowering effect of the insulin is opposed by its antagonist, and the glucose levels remain fairly constant. Since the blood glucose of Goldfine's (Goldfine et al., 1970) hypnotized patients did not change in spite of substantial increases in blood insulin, it would be interesting to know the composition of their imaginery meals, and whether circulating levels of glucagon can be altered by suggestion.

Metabolic Control Phases

I have tried to illustrate the point that Phase I of metabolic control—i.e., the initiation of signals by eating-associated phenomena—can be described as an intricate mix of substrate and hormonal signals arranged in sequential and parallel patterns (Figure 3-3). Although we do not know in detail the quantitative contribution of many of the components to the responses I have described, we can appreciate the extraordinary complexity, subtlety, and coordinate character of the responses. We can admire the appropriateness of the response to a meat meal, for example, as compared with that to a carbohydrate one, without necessarily understanding the mechanisms by which the response is achieved.

Phase II represents the delivery of information to cells in the form of hormones operating in a distinctive milieu of substrates (Figure 3-4). It cannot be emphasized too strongly that, in many ways, the substrates carry just as much directive information for involved cells as do the hormones. I am personally indebted to Randle and his colleagues (Newsholme and Randle, 1964), and their description of the glucose-fatty acid cycle, for emphasizing the importance of the metabolic mixture in the process of metabolic control. The hormones, by acting on a particular cell—e.g., as the adipocyte, may play an important role in determining the composition of the metabolic mixture offered to another cell downstream—e.g., the liver cell. As many people, including ourselves, have pointed out on many occasions, the extent of adipocyte

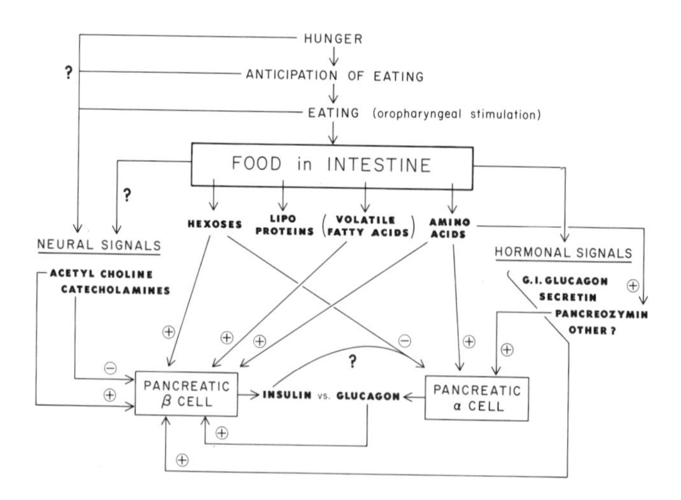

Figure 3-3. Schematic representation: metabolic control phases and signals.

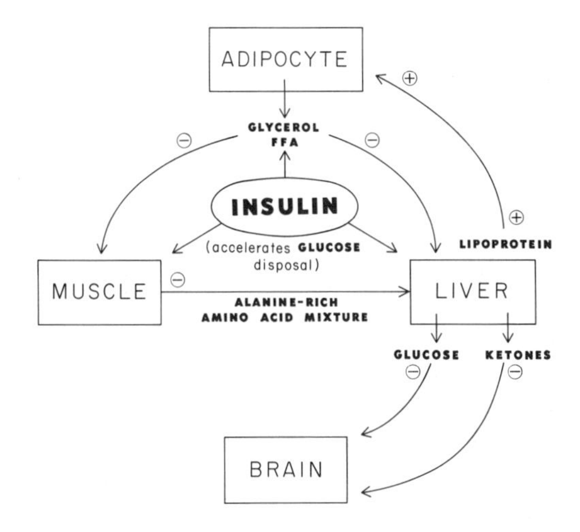

Figure 3-4. Schematic representation: Phase II of metabolic control.

lipolysis has profound effects on the metabolism of the hepatocyte. If ketogenesis and gluconeogenesis are not initiated primarily by fatty acids of depot origin, those processes are certainly sustained by such fatty acids (Toews et al., 1970). Conversely, if lipolysis in the adipocyte is effectively shut off, hepatic ketogenesis and gluconeogenesis are markedly inhibited. Coordinated with these shifts in the character of the perfusate are direct hormonal influences on the liver, principally those of insulin, glucagon, and glucocorticoids. It is possible to present a similar model for glucagon, and to demonstrate that the biochemical processes of glycogenesis, lipogenesis, and protein synthesis, all of which are stimulated by insulin, are inhibited by glucagon. On the other hand, gluconeogenesis and ketogenesis, which are stimulated by glucagon, are inhibited by insulin.

Phase III of metabolic control is the extension of the idea of a mix of intracellular signals—specialized "hormonal deputies" like 3′, 5′-cyclic AMP and substrates, metabolic intermediates, coenzymes, ions—generated by the arrival at the cell of the mix of hormonal and substrate signals we have been discussing. It is far beyond the scope of this chapter to attempt a detailed description of the kind of coordinate, patterned controls that operate in specific cells, such as the adipocyte and the hepatocyte. In their symmetry, their complexity, and their apparent appropriateness to the physiologic needs of the total organism they are very similar, at the cellular level of organization, to the more gross controls I have been discussing.

Phase IV of metabolic control is the imprinting of metabolic experience on the biochemical machinery of cells—the alteration in enzyme profile that occurs in cells required to perform specific metabolic chores over a period of time. For example, the brain "learns" how to use ketone bodies instead of glucose if food deprivation continues for a long enough time. The enzymatic "set" of the liver can accommodate either to total food deprivation or to continuous overfeeding (Tepperman and Tepperman, 1970). The kidney can alter its capacity to produce ammonia in the face of continued acidosis (Goldstein and Kensler, 1960). These long-range or "chronic" metabolic control mechanisms represent a sort of work hypertrophy or disuse atrophy of components of the biochemical machine. The patterns of enzyme adaptation are no less coordinate than are acute control mechanisms. They may, in fact, have their origin, in ways about which we can only speculate, in modified patterns of substrate flow initiated by acute control mechanisms.

Summary

Control mechanisms for carbohydrate and fat metabolism can be traced serially through the hierarchy of levels of biological organization from the molecule to behavior. However exciting contemplation of interacting molecules may be, it is most rewarding when they are seen as constituents of an intact, living, breathing, behaving organism of such astonishing complexity that we have only begun to understand the mechanisms by which it maintains its structual and functional integrity.

This work was made possible by Grant AM 5410 from the National Institute of Arthritis and Metabolic Diseases, U. S. Public Health Service. Thanks are due Walter Allan for permission to present original data, and Dr. Helen Tepperman for critical reading and valuable suggestions.

References

Allan, W., and Tepperman, H. M., 1969. Stimulation of insulin secretion in the rat by glucagon, secretin and pancreozymin: effect of aminophylline. *Life Sciences* 8:307.

Britton, S. W., 1925. Studies on the conditions of activity in endocrine glands. XVII. The nervous control of insulin release. *Am. J. Physiol.* 74:291.

Brobeck, J. W., 1939. The relation of hypothalamus to blood sugar and insulin sensitivity. Ph.D. diss., Northwestern University.

Coore, H. G., and Randle, P. J., 1964. Regulation of insulin secretion studied with pieces of rabbit pancreas incubated *in vitro. Biochem. J.* 93:66.

Daniel, P. M., and Henderson, J. R., 1967. The effect of vagal stimulation on plasma insulin and glucose levels in the baboon. *J. Physiol.* 192:317.

Esterhuizen, A. C., Spriggs, T. L. B., and Lever, J. D., 1968. Nature of islet-cell innervation in the cat pancreas. *Diabetes* 17:33.

Floyd, J. C., Jr., Fajans, S. S., Conn, J. W., Knopf, R. F., and Rull, J., 1966. Insulin secretion in response to protein ingestion. *J. Clin. Invest.* 45:1479.

Frohman, L. A., Esdinli, E. Z., and Javid, R., 1966. Effect of vagal stimulation on insulin secretion. *Diabetes* 15:522.

Goldfine, I. D., Ryan, W. G., and Schwartz, T. B., 1969. Effect of glucola, diet cola and water ingestion on blood glucose and plasma insulin. *Proc. Soc. Exptl. Biol. and Med.* 131:329.

Goldfine, I. D., Abraira, C., Gruenewald, D., and Goldstein, M. S., 1970. Plasma insulin levels during imaginary food ingestion under hypnosis. *Proc. Soc. Exptl. Biol. and Med.* 133:274.

Goldstein, L., and Kensler, C. J., 1960. Factors which affect the activity of glutaminase I in the guinea pig kidney. *J. Biol. Chem.* 235:1086.

Grey, N., McGuigan, J. E., and Kipnis, D. M., 1970. Neutralization of endogenous glucagon by high titer glucagon antiserum. *Endocrinology* 86:1383.

Han, P. W., Yu, Y-K., and Chow, S. L., 1970. Enlarged pancreatic islets of tube-fed hyposectomized rats bearing hypothalamic lesions. *Am. J. Physiol.* 218:769.

Hinkle, L. E., Jr., Evans, F. M., and Wolf, S., 1951. Studies in diabetes mellitus. III. Life history of three persons with labile diabetes, and relation of significant experiences in their lives to the onset and course of the disease. *Psychosom. Med.* 13:16.

Horino, M., Machlin, L. J., Hertelendy, F., and Kipnis, D. M., 1968. Effect of short chain fatty acids on plasma insulin in ruminants and nonruminant species. *Endocrinology* 83:118.

Kaneto, A., Kosaka, K., and Nakao, K., 1967. Effects of stimulation of the vagus nerve in insulin secretion. *Endocrinology* 80:530.

Kraegen, F. W., Chisholm, D. J., Young, J. D., and Lazarus, L., 1970. The gastrointestinal stimulus to insulin release. II. A dual action of secretin. *J. Clin. Invest.* 49:524.

Malaisse, W., Malaisse-Lagae, F., Wright, P. H., and Ashmore, J., 1967. Effects of adrenergic and cholinergic agents upon insulin secretion *in vitro*. *Endocrinology* 80:975.

Manns, J. G., Boda, J. M., and Wiles, R. F., 1967. Probable role of propionate and butyrate in control of insulin secretion in sheep. *Am. J. Physiol.* 212:756.

Mason, J. W., 1968. Organization of psychoendocrine mechanisms. *Psychosom. Med.* 30:565.

Mayhew, D. A., Wright, P., and Ashmore, J., 1969. Regulation of insulin secretion. *Pharmacol. Rev.* 21:183.

McIntyre, N., Holdsworth, C. D., and Turner, D. S., 1965. Intestinal factors in the control of insulin secretion. *J. Clin. Endocrinol. Metab.* 25:1317.

Muller, W. A., Faloona, G. R., Aguilar-Parada, E., and Unger, R. H., 1970. Abnormal alpha-cell function in diabetes. *New England J. Med.* 283:109.

Newsholme, E. A., and Randle, P. J., 1964. Regulation of glucose uptake by muscle. 7. Effects of fatty acids, ketone bodies and pyruvate, and of alloxan-diabetes, starvation, hypophysectomy, and adrenalectomy, on the concentrations of hexose phosphates, nucleotides and inorganic phosphate in perfused rat heart. *Biochem. J.* 93:641.

Porte, D., Jr., 1969. Sympathetic regulation of insulin secretion. *Arch. Intern. Med.* 123:252.

Samols, E., Marri, G., and Marks, V., 1965. Promotion of insulin secretion by glucagon. *The Lancet* 2:415.

Stern, J. S., Baile, C. A., and Mayer, J., 1970. Are propionate and butyrate physiological regulators of plasma insulin in ruminants? *Am. J. Physiol.* 219:84.

Sussman, K. E., Vaughan, G. D., and Timmer, R. F., 1966. Factors controlling insulin secretion from the perfused isolated rat pancreas. *Diabetes* 15:521.

Tepperman, J., and Tepperman, H. M., 1970. Gluconeogenesis, lipogenesis and the Sherringtonian metaphor. *Fed. Proc.* 29:1284.

Toews, C. J., Lowy, C., and Ruderman, N. B., 1970. The regulation of gluconeogenesis. *J. Biol. Chem.* 245:818.

Unger, R. H., Ketterer, H., Dupre, J., and Eisentraut, A. M., 1967. The effects of secretin, pancreozymin, and gastrin on insulin and glucagon secretion in anesthetized dogs. *J. Clin. Invest.* 46:630.

Williams, R. H., and Ensinck, J. W., 1966. Secretion, fates and actions of insulin and related products. *Diabetes* 15:623.

4

Control Mechanisms in Amino Acid Metabolism

Alfred E. Harper

The term "control mechanisms" carries the connotation of "the molecular basis of regulation." While we know something of the molecular basis of regulation of amino acid metabolism in higher organisms, we are far from being able to present an integrated panorama with the detail of a sharply focused photograph. So, rather than concentrate on the details that presently appear reasonably clear, I have chosen to paint with the brush of an impressionist a general view of regulation of amino acid metabolism, with the fine detail purposely left incomplete.

Homeostasis

Despite all the modern terminology derived from the age of servomechanisms, when we talk about regulation and feedback mechanisms we are talking about a basic biological concept developed long before the age of electronics. This is the concept of homeostasis. Claude Bernard (1878), during the latter half on the nineteenth century, recognized that homothermic animals were little incapacitated by drastic environmental changes because they lived essentially in their own internal environment which remained quite constant despite fluctuations in the external environment. But more importantly, he recognized that this was not because the organism was separated from the external environment but rather that it was responsive to external changes, for which it continuously and delicately compensated. He included the supply of nutrients among the components of the external environment that could fluctuate greatly.

The concept was extended by Henderson (1928), who was impressed as much by the large fluctuations that occurred in the concentrations of blood constituents under various conditions as he was by the constancy of fasting values. The ability of the body to restore the blood to its standard state led him to recognize the phenomenon of adaptation as a basic biological phenomenon. He also recognized that such mechanisms had limits, that with too great an external influence the body could not restore the standard state and would have to compromise at something less than the ideal. "Adaptation," he stated, "is relative. It involves a question, not of what is best, but of what is efficient under certain conditions, of what promotes survival in a particular environment . . . . "

Cannon (1929) proposed the term "homeostasis" to describe Bernard's concept. He was particularly concerned with the problem of the integration of regulatory mechanisms within the complex animal body to ensure stability. He applied the feedback concept to regulation of a system in the steady state. "The highly developed living being," he wrote, "is an open system having many relations to its surroundings . . . . changes in the surroundings excite reactions in this system or affect it directly, so that internal disturbances of the system are produced. Such disturbances are normally kept within narrow limits, because automatic adjustments within the system are brought into action . . . . "

Development of increasingly effective systems for the maintenance of homeostasis was emphasized by Sir Joseph Barcroft (1932) as the mechanism of evolution. He was disturbed, however, to note that many "lower" animals could tolerate greater deviations in blood constituents and still remain functional, then could "higher" organisms. But in examining such phenomena he noted that the initial effects of deviations of various blood constituents beyond the "normal" range caused loss of mental coordination. He concluded that higher mental activity and intellectual development, which permit the greatest possible freedom from the external environment, depended upon an accuracy of control of the internal environment that could be achieved only with complex, integrated control mechanisms. An example is the occurrence of mental defects in man when control mechanisms for the regulation of tissue amino acid concentrations are ineffective, owing to genetic deletion of enzymes of amino acid catabolism.

For accurate regulation of amino acid metabolism, body fluid amino acid concentrations must be monitored by a system that can detect deviations from some standard value; and that, when a deviation occurs, can set into motion an action that will tend to restore the original state. Evidence for the existence of such a system or systems is supported by observations that plasma amino acid concentrations do not fall greatly during a prolonged fast (Adibi, 1968), and that amino acids are cleared rapidly from the blood after a large load of an amino acid has been ingested or injected (Coulson and Hernandez, 1968).

Regulatory Mechanisms

Changes in the rates of enzymatic reactions and changes in the rates of transport of nutrients and metabolic products are two regulatory mechanisms common to all biological systems. The rate of an enzymatic reaction is determined by the concentration of substrate, of cofactor, and of any activator or inhibitor that may be present in the cell. These factors influence enzymatic activity regardless of the enzyme content of the organism. Besides these influences of small molecules, the amount of enzyme may increase or decrease in response to a change in the nutritional or endocrine state of the organism. Enzymatic adaptations of this type may greatly alter the capacity of a metabolic system. Also, the rate of transport of nutrients is determined by the concentration of the nutrient in the extracellular fluid. If transport is carrier-mediated it may also be influenced by the concentration of any cofactors, activators, or inhibitors in the system. Much less is known about the characteristics of transport systems and the factors influencing them than about enzymes, but there is evidence that, in microorganisms, the amount of carrier may change in response to changes in nutrient supply (Kaback, 1970) and that, in higher organisms, hormones may influence the rate of transport by altering the capacity of transport systems.

In higher organisms, with their great complexity, systems are required also for the integration and coordination of metabolism. The endocrine system and the central nervous system provide for this by their ability to alter the rates of enzyme activity and transport differentially in various organs and tissues.

When we talk about control mechanisms in amino acid metabolism, what are we concerned with? First and foremost, with the trapping and conservation of the indispensable or essential amino acids that must be obtained from the external environment for the synthesis of tissue proteins to ensure growth or, at the very least, survival. Secondly, we are concerned with the disposal of surpluses of amino acids beyond the amounts needed for growth or maintenance, as surpluses of amino acids can be deleterious, even toxic, and must be degraded and the degradation products eliminated to ensure survival. If the external environment provided constantly just the amounts of amino acids needed by the body there would be little need for elaborate regulatory mechanisms.

A schematic outline identifying possible sites of regulation of amino acid metabolism is shown in Figure 4-1. The initial possibility is regulation in the gastrointestinal tract of the rate of entry of amino acids into the blood; after entry into the blood, amino acids flow through the portal vein to the liver, where both protein synthesis and amino acid degradation are major metabolic processes. Amino acids that are not removed

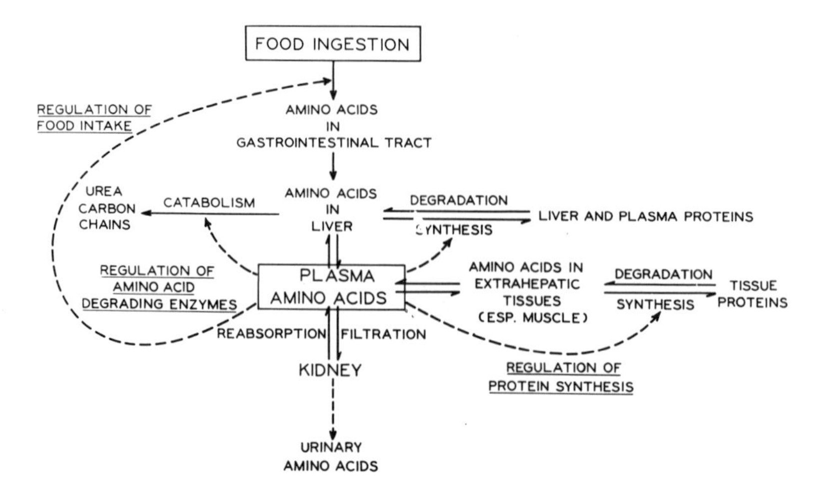

Figure 4-1. Schematic representation of sites of regulation of amino acid metabolism that may be influenced by plasma amino acid concentrations. (Harper and Benevenga, 1970)

by the liver pass, via the circulatory system, to other visceral organs and to the peripheral tissues, where protein synthesis is the major activity, but where metabolism of the dispensable amino acids and the branched-chain amino acids is also an important metabolic activity. The possibility of regulation in relation to conservation or excretion arises again as amino acids circulate through the kidney. Besides the possibilities of regulation of transport during absorption, entry into tissues, and urine formation in the kidney, and regulation of incorporation into proteins and of degradation, mainly in the liver, there is as well regulation of protein or amino acid intake—conceivably, as a component of amino acid transport—which also represents a gross system for the control of metabolism.

Gastrointestinal Tract as a Site of Regulation

Regulation of the metabolism of nutrients begins in the gastrointestinal tract (Munro, 1964a). The rate of passage of nutrients into the circulatory system depends upon the rate of gastric emptying, which is influenced by many factors, one of which is the amount of protein consumed. The lower curve (A) in Figure 4-2 shows the steady, smooth emptying of a protein-free meal from the stomachs of rats fed 5 gm of diet (Peraino et al., 1959). The upper curves (B, C, D) show that emptying is delayed when the protein content of the diet is increased. Regulation of stomach emptying contributes to the conservation of dietary amino acids by reducing the possibility of excretory losses that can occur when the intestine is loaded beyond its capacity for digestion.

The amino acids arising from protein digestion are absorbed actively from the lumen of the intestine. The intestinal mucosa has evolved as an immense absorptive surface owing to its protruding villae and microvillae. The enormous capacity for absorption is illustrated in Figure 4-3, which shows that, in the rat, the rate of nitrogen disappearance increases threefold, even though gastric emptying is delayed, when the protein content of the diet is increased from 15 to 50%, and that this rate remains steady over a 6-hour period (Peraino et al., 1959). The overall efficiency of the system is evident from the observation that the digestibility of wheat gluten was unaltered when the dietary protein content was increased from 30 to 70% (Munaver and Harper, 1959). Further, even when toxic amounts of amino acids are ingested with a low-protein diet, negligible amounts are excreted in the feces (Figure 4-4) as evidenced by the small amount of radioactivity excreted by rats fed a diet containing 5% of C^{14}-labeled L-leucine (Tannous, 1963). This is evidence that regulation of intestinal absorption of amino acids does not contribute to the regulation of amino acid metabolism. Despite the basic high capacity of the intestine for absorption, rats fed a high-protein diet undergo adaptation that increases further the capacity of the intestine to absorb amino acids (Scharrer et al., 1967).

The gastrointestinal tract thus functions as a retrieval system for amino acids that are present in the proteins of digestive secretions (Rogers and Harper, 1966). Also, the rate and extent of amino acid ox-

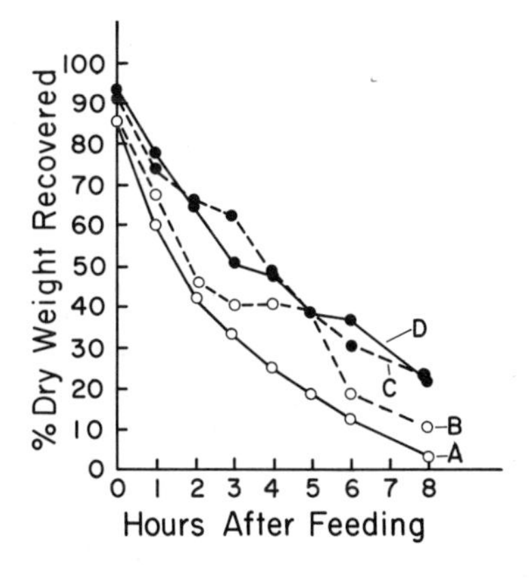

Figure 4-2. Effect of protein content of diet on speed of stomach emptying in the rat. Curve A, protein-free diet; B, 15% casein diet; C, 30% casein diet; and D, 50% casein diet. (After Peraino et al., 1959)

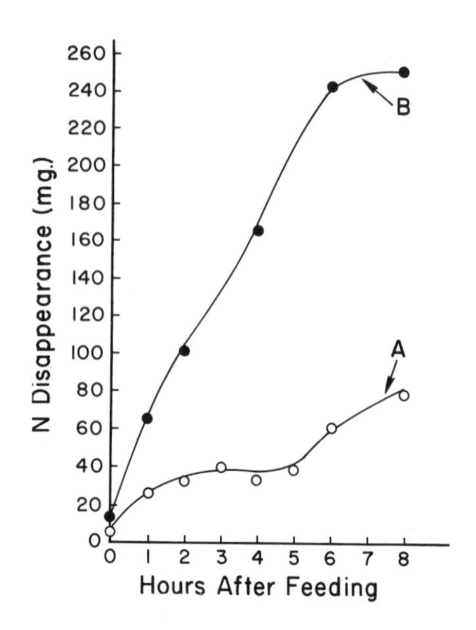

Figure 4-3. Disappearance of nitrogen from gastrointestinal tract of rats fed different quantities of protein. Curve A, 15% casein diet; Curve B, 50% casein diet. (After Peraino et al., 1959)

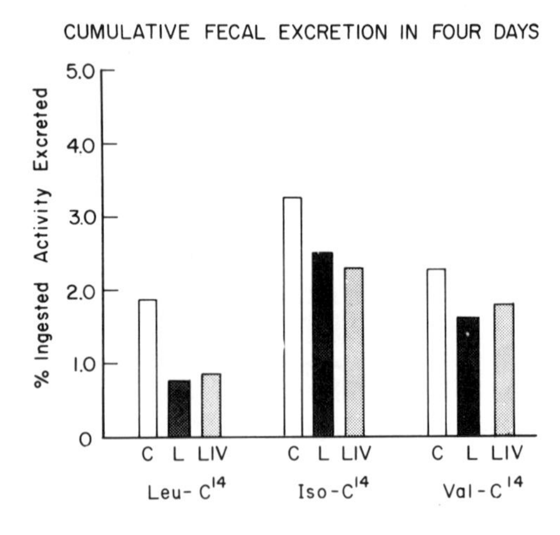

Figure 4-4. Cumulative fecal excretion of radioactivity over 4 days by rats fed 9% casein (C), high-leucine (L), and high-leucine with isoleucine and valine (LIV) diets together with C^{14}-labeled leucine, isoleucine, or valine. (Tannous, 1963)

idation depends upon the concentration of amino acids in body fluids; control of the rate of flow of amino acids to sites of protein synthesis through regulation of gastric emptying should reduce the probability of protein-synthesizing systems being overloaded and should thereby tend to minimize losses of amino acids through degradative reactions.

Amino Acid Transport

After being absorbed, amino acids enter the portal blood which flows to the liver. Those not utilized by the liver pass into the systemic circulation and to other organs and tissues. As the rate of metabolism can be influenced by the supply of substrate, the rate of transport can become a controlling factor in metabolism; therefore, mechanisms that facilitate entry of amino acids into cells would seem essential. If the ratio of the concentration of an amino acid in a tissue to that in blood exceeds 1, it is highly probably that the amino acid is transported across the cell membrane by an active process. Ratios greater than 1 are observed regularly for most amino acids in animal tissues (Table 4-1) (Tallan et al., 1954).

It has been well established by in-vitro transport studies that the rate of transport of a solute by a carrier-mediated system increases with increasing concentration of the solute until the carrier is saturated. This is illustrated for the model amino acid, α-aminoisobutyric acid (AIB), in liver slices, in Figure 4-5 (Tews and Harper, 1969). The rate of entry of an amino acid into tissues should therefore increase when the concentration in blood increases after a meal (McLaughlan and Illman, 1967).

Table 4-1. Concentrations of Some Amino Acids in Tissues of the Cat (mg/100 g tissue; mg/100 ml plasma)

Amino acid	*Plasma*	*Liver*	*Brain*	*Muscle*	*Kidney*
Threonine	1.4	3.1	2.6	3.9	3.6
Leucine	1.6	3.6	1.8	2.3	3.2
Lysine	2.8	3.6	2.0	5.5	3.7
Glycine	2.3	9.1	10.1	6.7	14.4
Glutamic Acid	1.8	66.0	128.0	36.0	127.0

(After Tallan et al., 1954.)

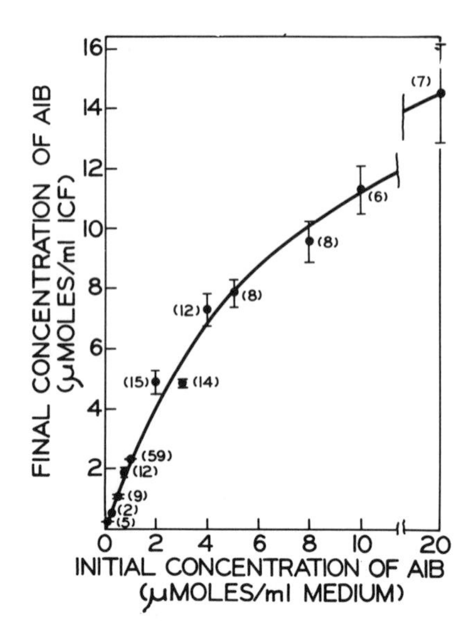

Figure 4-5. Effect of concentration of α-aminoisobutyric acid (AIB) in incubation medium on transport of AIB into liver slices, expressed as the uptake of AIB in moles per ml of intracellular fluid (ICF). (Tews and Harper, 1969)

Amino acid uptake by muscle is stimulated by insulin, as also is amino acid incorporation into muscle protein (Wool et al., 1968). As insulin secretion is stimulated by an influx of glucose into the blood, uptake of amino acids by muscle should increase whenever a meal is ingested (Munro, 1964b). Amino acid uptake by the isolated perfused liver (Mallette et al., 1969) and by liver slices (Tews et al., 1970) is increased by glucagon and AIB uptake is greater in liver slices from rats fed a high-protein diet than in slices from rats fed a low-protein diet (Figure 4-6). Cyclic-3′, 5′-adenosine monophosphate (c-AMP) concentration is elevated in liver from rats fed a high-protein diet (Jost et al., 1970) and AIB uptake by liver slices is stimulated by c-AMP (Tews et al., 1970).

Far-reaching conclusions about these relationships should not be drawn until the evidence is more complete, but that presently available suggests (1) that secretion of both insulin and glucagon is stimulated when protein intake is high; (2) that the rate of removal of surpluses of amino acids and their utilization for gluconeogenesis by liver is increased by glucagon; and (3) that uptake of amino acids into muscle and their incorporation into muscle proteins is stimulated by insulin. This leads to the view that regula-

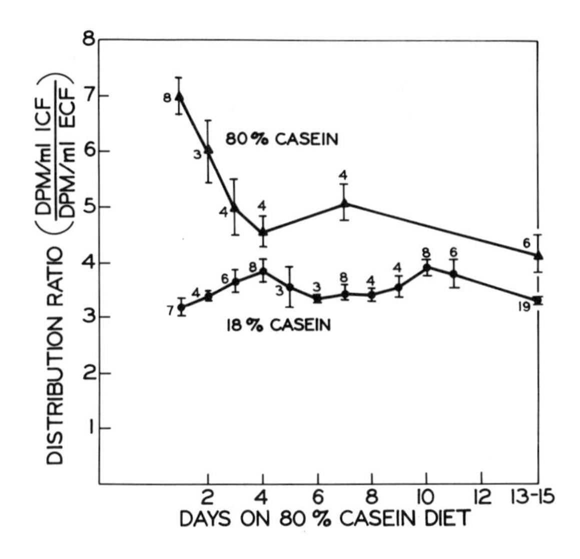

Figure 4-6. Transport of α-aminoisobutyric acid by liver slices from rats fed 18% or 80% casein diets, expressed as the ratio of radioactivity (disintegrations per minute, dpm) in the intracellular (ICF) and extracellular (ECF) fluids. (Tews et al., 1970)

tion of transport through alterations in solute concentration and regulation by endocrine secretions function to conserve amino acids for tissue protein synthesis when the supply is limited and to remove amino acids which might otherwise accumulate and lead to adverse effects when the amount consumed is excessive.

In discussing possible sites of regulation of amino acid metabolism by transport mechanisms, reabsorption from and excretion by the kidney should also be considered. The kidney is a major organ for homeostatic regulation of osmotic pressure and pH of body fluids, and for removal of waste products and toxic substances. The kidney tubule, like most other tissues, has mechanisms for active transport of amino acids; the low loss of amino acids in the urine of animals fed a high-protein diet is evidence that reabsorption of amino acids by the kidney is highly efficient. The amount of radioactivity excreted in urine by rats fed a high-leucine diet containing C^{14}-leucine (Tannous, 1963) is small (Figure 4-7) even when the amount ingested is high enough to cause adverse effects. The kidney serves as an organ of excretion only when the renal threshold is exceeded. When rats are fed a diet containing 5% of tyrosine they become severely debili

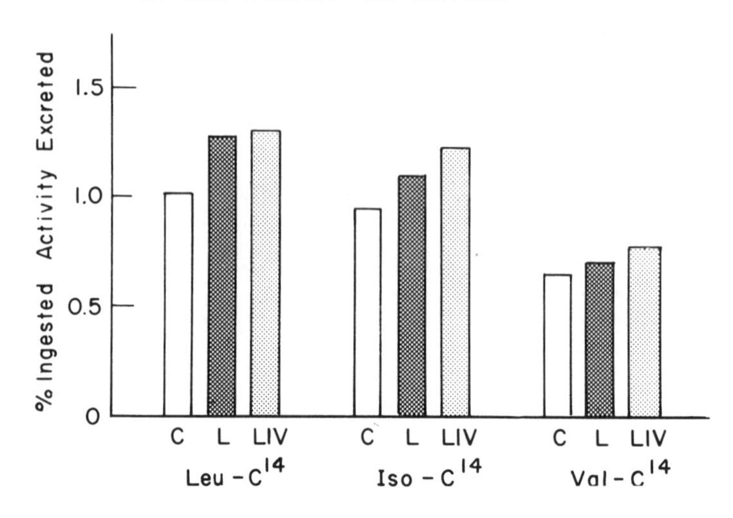

Figure 4-7. Urinary excretion of radioactivity over 24 hours by rats fed 9% casein (C), high-leucine (L), and high-leucine with isoleucine and valine (LIV) diets together with C^{14}-labeled leucine, isoleucine, or valine. (Tannous, 1963)

tated, and although tyrosine is excreted in the urine, plasma tyrosine remains greatly elevated (Boctor, 1967). A similar situation appears to hold in rats fed a diet devoid of one indispensable amino acid (Pearce et al., 1947). Thus, excretion of amino acids in urine does not serve as a mechanism for elimination of an excess of amino acids from body fluids; evidently such elimination occurs only when amino acid concentrations are in the range associated with adverse effects.

Kidney reabsorption thus represents a mechanism for the retrieval of amino acids and like the intestine functions as a unidirectional trapping system.

As has the intestine, the kidney has evolved as an organ that functions as a highly efficient mechanism for conservation of amino acids.

Amino Acid Pools; Lack of Storage Mechanism

A major regulatory mechanism for the removal of surpluses of energy substrates is storage in forms that can be mobilized subsequently during periods of deprivation.

Table 4-2. Comparison of Amino Acid Pools and Requirements of Rat (μmoles/100 g body weight)

Amino acid	*Plasma*	*Intestine*	*Liver*	*Muscle*	*Requirement*
Threonine	1.1	2.8	1.6	72	400
Lysine	1.1	3.9	1.7	32	600
Histidine	0.3	1.2	3.1	33	140
Valine	0.8	3.4	1.5	8	500
Leucine	0.5	5.7	2.6	6	500

(After Munro, 1970)

The amounts of free amino acids in organ and tissue pools as shown in Table 4-2 are negligible in relation to amino acid requirements (Munro, 1970). Also, the body protein content of animals fed a high-protein diet is very little greater than that of controls fed a diet that just meets the requirement for protein (Mayer and Vitale, 1957). Thus storage of amino acids either as protein or as free amino acids provides little in the way of amino acid reserves. In fact, a portion of the increase in protein content that is observed in animals fed a high-protein diet is associated with hypertrophy of organs such the liver and kidney and represents adaptation to the overloading of these organs rather than a mechanism for amino acid conservation.

Further evidence of the lack of amino acids stores comes from studies of the effects of delayed amino acid supplementation. In these studies, animals are fed a diet that is complete except for one amino acid, and the missing amino acid is fed subsequently (Elman, 1939). If the missing amino acid is provided even a few hours later growth is retarded, but the severity of the effect depends upon which amino acid has been omitted. Delayed supplementation of lysine retards growth much less than delayed supplemention of isoleucine (Spolter and Harper, 1961). The rapid rate of disappearance of radioactivity from the liver of rats injected with C^{14}-lysine provides direct evidence of the short survival time of free amino acids (Haider and Tarver, 1969). Also, Black (1968) has shown that the half-lives of amino acids infused into the cow are only a few minutes (Figure 4-8).

Thus, although storage represents an important regulatory mechanism for the disposal of a surplus of energy it does not function as a regulatory mechanism for removal of a surplus of amino acids.

Nevertheless, tissue proteins are mobilized when amino acid intake is inadequate. This is the result of the normal process of protein turnover,

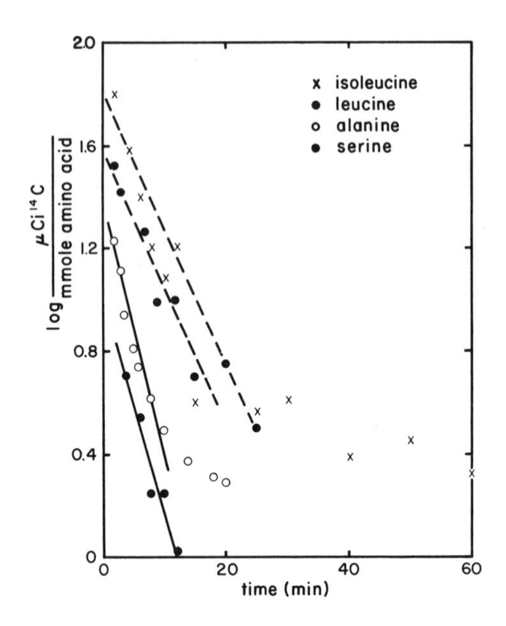

Figure 4-8. Changes in specific activity of plasma amino acids over time after intravenous injection of dairy cows with uniformly C^{14}-labeled L-amino acids. (Black, 1968)

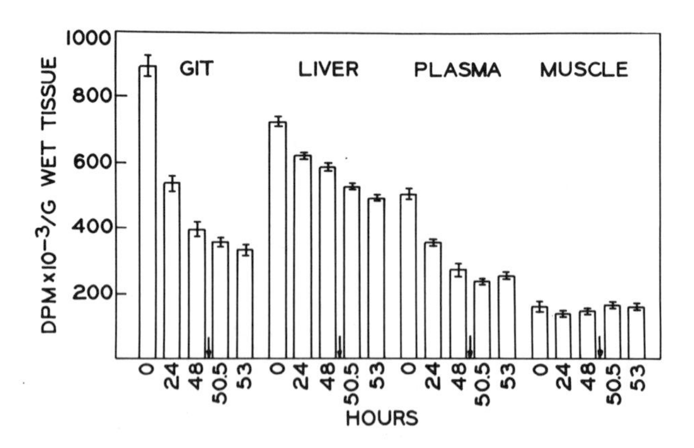

Figure 4-9. Changes in radioactivity of protein fractions of tissues and blood plasma over time after feeding rats uniformly C^{14}-labeled L-lysine. (Soliman and Harper, 1969)

with rapidly turning-over proteins, especially in the intestinal mucosa and liver, being depleted initially and the more slowly turning-over proteins, especially in muscle, being depleted subsequently. Turnover in nondepleted rats is illustrated by observations, summarized in Figure 4-9, from an experiment in which disappearance of C^{14}-lysine from tissue proteins of rats previously labeled with radioactive lysine was followed for two days and then for a few hours after a meal. Radioactivity disappeared rapidly from the gastrointestinal tract, more slowly from liver, and not at all from muscle under these conditions (Soliman and Harper, 1969).

The proteins depleted during periods of protein deprivation are not storage proteins; they are functional tissue proteins. Also, the process is evidently regulated, as amino acid-degrading enzymes which are depleted severely in rats fed a protein-deficient diet are not depleted during starvation (Harper, 1965). As well, the rate of turnover of both serum and liver proteins may be prolonged (Table 4-3) and reutilization of amino acids for resynthesis may be increased during periods of protein deprivation (Waterlow, 1968). These responses should contribute to survival during periods of deprivation: enzymes required for gluconeogenesis are maintained at the expense of other proteins when the primary need, as in starvation, is glucose for the central nervous system; and enzymes for

Table 4-3. Oxidation of C^{14}-Labeled Amino Acids by the Perfused Liver and by the Eviscerated Carcass

Amino acid	*Administered*	*C^{14} expired as $C^{14}O_2$ (%)*
	Eviscerated surviving rat	*Isolated perfused liver*
L-phenylalanine	0.14	25.5
L-histidine	0.44	30.0
DL-tryptophan	1.7	9.1
L-lysine	2.4	34.5
L-isoleucine	7.7	5.2
L-leucine	12.0	5.4
Glycine	7.1	19.0
L-Glutamic acid	46.0	32.0

(After Miller et al., 1956)

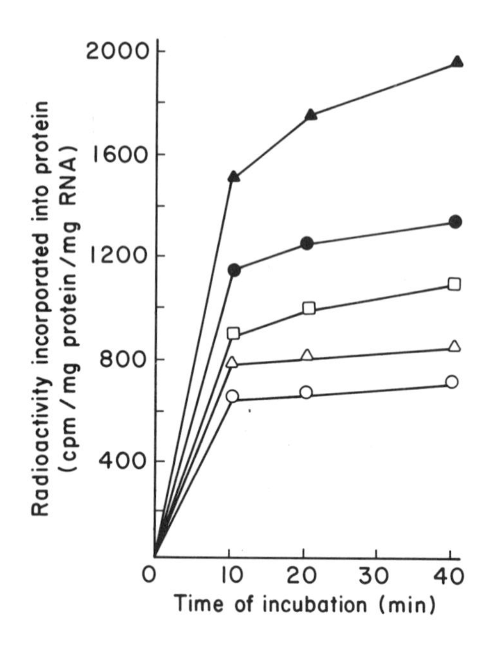

Figure 4-10. Amino acid incorporation by a cell-free system prepared from rat livers infused for 60 minutes with media containing different amounts of amino acids. Open circles, no added amino acids; open triangles, plasma concentrations; squares, three times plasma concentrations; filled circles, five times plasma concentrations; filled triangles, ten times plasma concentrations. (Jefferson and Korner, 1969)

amino acid degradation fall substantially during protein depletion when the primary need is for amino acid conservation.

Synthesis of Proteins

Although the evidence comes from microbial systems, the Michaelis constants (K_m) for amino acids in protein-synthesizing systems are well below those for most amino acid-degrading systems, a mechanism which should ensure that first priority is given to protein synthesis when amino acid supply is low.

Evidence that protein-synthesizing systems are responsive to the supply of amino acids has been accumulating, especially from the work of Munro (1968) and Sidransky and associates (1968). The basis for this response has not been established, but a supply of amino acids in some way stimulates aggregation of ribosomes into polysomes. This aggregation is inhibited if starved rats are subsequently fed a diet from which tryptophan is deleted, but if the amino acid pool of liver is depleted of other amino acids, aggregation is dependent on the supply of amino acids other than tryptophan. The integrity of subcellular structures involved in protein synthesis thus depends upon the dietary supply of amino acids. Jefferson and Korner (1969) have reported that the capacity of the perfused liver to synthesize protein is a function of the amino acid supply (Figure 4-10).

The rate of protein synthesis in muscle falls rapidly in protein-deficient rats (Waterlow, 1968). Also, the activities of amino acid-activating enzymes increase in animals fed a protein-deficient diet (Gaetani et al., 1964), a response that should lead to increased efficiency of trapping of amino acids for protein synthesis when the supply of amino acids is inadequate. This mechanism, together with the rapid reaggregation of ribosomes when amino acids and energy are supplied, may well account for the rapid "catch-up" growth observed when protein-deficient animals are provided with an adequate amount of protein.

Catabolism of Amino Acids

Amino acids continue to be absorbed by the intestine and to be conserved by the kidney even after the amounts required for protein synthesis have been exceeded. Since amino acids are not stored and since excessive amounts of both amino acids and of ammonia, a major end-product of amino acid metabolism, are toxic, any excess of amino acids must be removed through degradation and elimination of the degradation products. Also, amino acid catabolism provides the building blocks for glucose synthesis via gluconeogenesis when the supply of glucose for the functioning of the central nervous system is inadequate. With such great dependence on this pathway for disposal of extra amino acids, efficient regulation of it would be predicted.

It is instructive to look first at the distribution of the enzymes of amino acid degradation. One rather general approach is by comparing the ability of the isolated perfused liver and the eviscerated surviving carcass to degrade individual amino acids (Table 4-4), as was done by Miller et al. (1956). The indispensable amino acids, except for the branched-chain amino acids, are degraded almost exclusively in the liver. This type of enzyme distribution should contribute to conservation of indispensable amino acids that pass through the liver for the synthesis

Table 4-4. Half-Lives of Rat Serum and Liver Proteins, as Measured with C^{14}-Arginine (days)

Proteins	*True*	*Apparent*
Serum mixed		
Normal diet	2.0	2.7
Low-protein diet	3.3	5.5
Liver		
Normal diet	1.9	5.5
Low-protein diet	2.3	9.2

(After Waterlow, 1968.)

of proteins in the peripheral tissues. Here is an evolutionary development with survival value. Use of amino acids primarily for protein synthesis in the liver is favored by another evolutionary development. The Michaelis constants of amino acid-activating enzymes for amino acids are low, so protein-synthesizing systems should be saturated even when amino acid concentrations are low, whereas those for amino acid-degrading enzymes are so high that they probably function only rarely at their maximum rate.

The metabolism of the branched-chain amino acids is unique, in that the first step, transamination, is highly active in kidney and peripheral tissues while oxidation of the keto acids formed is much more active in the liver than in any other tissue (Harper et al., 1970). The explanation is not obvious on a teleological basis. However, leucine and isoleucine are effective stimulators of insulin secretion, which in turn stimulates uptake of amino acids by muscle and also stimulates amino acid incorporation. It is thus possible that passage of these amino acids through the liver unaltered also contributes to amino acid conservation by stimulating more efficient utilization of all amino acids for tissue formation in the periphery.

Most of the enzymes for degradation of the dispensable amino acids are widely distributed. Since the dispensable amino acids can be synthesized by the body from carbon skeletons derived from glucose, to maintain an adequate supply of them requires primarily the conservation of nitrogen. Wide distribution of glutamate dehydrogenase and glutamine synthetase ensures effective trapping of ammonia or amino groups released in cells through synthesis of glutamate and glutamine as long as there is a supply of α-ketoglutarate. These two compounds can then serve as precursors for the synthesis of other dispensable amino acids. Glutamate dehydrogenase, as well as being widely distributed, is a highly active enzyme in liver and kidney, and its equilibrium constant is such that formation of glutamate rather than its degradation is favored. Glutamate-oxaloacetate and glutamate-pyruvate transaminases are also widely distributed and contribute to the redistribution of nitrogen.

The activities of most of the enzymes for degradation of the indispensable amino acids are relatively low in animals fed normal diets, but because of the high K_m's of these enzymes the rate of the reaction increases rapidly as the concentration of substrate increases. Kim and Miller (1969) have reported that an increase in tyrosine concentration in the perfusion medium results in a substantial increase in tyrosine degradation by the isolated perfused liver. Also, with rats fed a high-methionine diet containing methionine-l-C^{14} the amount of expired $C^{14}O_2$ increases as the amount of methionine ingested is increased (Benevenga and Harper, 1970).

Although the rate of amino acid degradation increases when the concentrations of amino acids in body fluids increase, greatly elevated plasma amino acid concentrations are observed when animals are fed large loads of individual amino acids, particularly if the animal has previously been fed a low-protein diet (Sauberlich, 1961). This observation indicates that amino acid-degrading capacity can be exceeded even though the rate of amino acid degradation responds when amino acid concentrations in body fluids increase.

Another type of mechanism for regulation of amino acid degradation is evident from the response of tryptophan oxygenase in animals administered a load of trypyophan (Knox, 1963). Here, tryptophan increases the affinity of the enzyme for its cofactor, with a resulting increase in enzyme activity. Tryptophan administration also prevents degradation of the enzyme protein (Schimke et al., 1965); thus the activity of the enzyme gradually builds up further. Relatively few amino acid-degrading enzymes respond in animals ingesting large quantities of individual amino acids, but many respond when total protein intake is increased. It seems improbable that conditions leading to selection of mutations for the responsiveness of individual amino acid-degrading enzymes to a load of their substrates would have existed in nature, so it is not surprising that this phenomenon is rather rare. On the other hand, conditions leading to selection for more general adaptive responses of amino acid-degrading enzymes—e.g., to starvation or a high-protein intake—would not be unexpected and a large number of such responses are known (Schimke and Doyle, 1970).

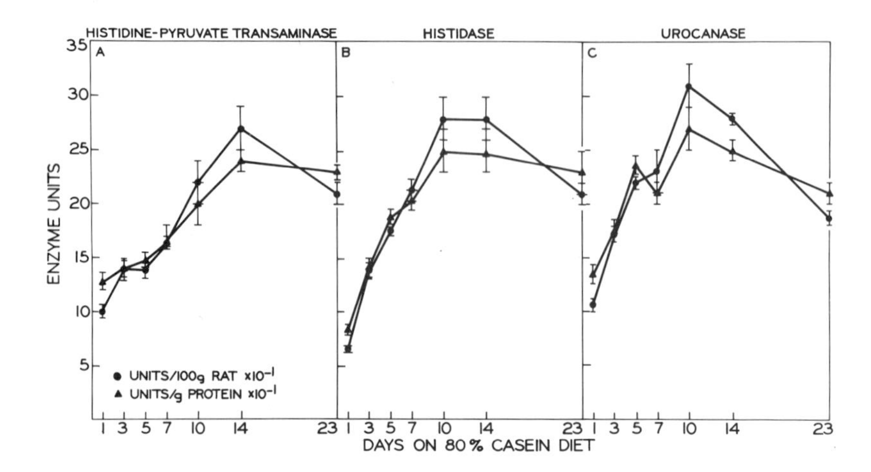

Figure 4-11. Changes in activities of histidine-degrading enzymes over time in livers of rats fed an 80% casein diet. (Schirmer and Harper, 1970)

Threonine dehydratase shows only small responses to increases in protein intake until the amount of protein ingested exceeds the requirement; then, very substantial increases are observed (Harper, 1968). Tyrosine transaminase shows a similar pattern (Rosen et al., 1959). Histidase responds markedly as protein intake increases (Schirmer and Harper, 1970), as also do all of the enzymes of the urea cycle (Schimke, 1962, 1963). The pattern of response with time to a high-protein intake is shown for the histidine-degrading enzymes in Figure 4-11, and the rapid rate of return to a low value after protein is removed from the diet is shown in Figure 4-12.

The mechanism involved is not established for each enzyme, but both stimulation of enzyme synthesis and inhibition of enzyme degradation have been demonstrated under different conditions, along with evidence that the rate of response is a function of the rate of turnover of the enzyme and the amount of enzyme present, the response being evident very quickly with enzymes having a rapid rate of turnover (Schimke and Doyle, 1970).

Interestingly, the branched-chain amino acid transaminases appear to show few adaptive responses, but the branched-chain keto acid dehydrogenases respond over the lower range of protein intakes in animals fed different amounts of protein; in particular, they fall to quite low activities when protein intake is low. Since transamination is a readily reversible process, this mechanism could serve for the conservation of these amino acids when protein intake is low (Wohlhueter and Harper, 1970).

Many of the amino acid-degrading enzymes also increase in activity in animals treated with glucocorticoids or glucagon, hormones known to stimulate gluconeogenesis. The hormonal treatments appear to act primarily through stimulating enzyme synthesis (Schimke and Doyle, 1970).

Amino acid-degrading capacity is thus responsive to changes in protein or amino acid intake. A decrease in degradative capacity when protein intake is low should contribute to conservation of amino acids for protein synthesis. An increase in degradative capacity when protein intake is high should increase the rate of removal of amino acids and stimulate gluconeogenesis. As shown in Figure 4-13, the rate of disappearance of histidine from the perfusion medium is increased when the liver undergoing perfusion is from a rat in which histidine-degrading enzymes have been induced (Mossie and Harper, unpublished data). Responses such as these are indicative of effective homeostatic regulation of amino acid degradation.

Food Intake as a Regulatory Mechanism

With no capacity for amino acid storage nor any mechanism for preventing absorption from the intestine or reabsorption from the

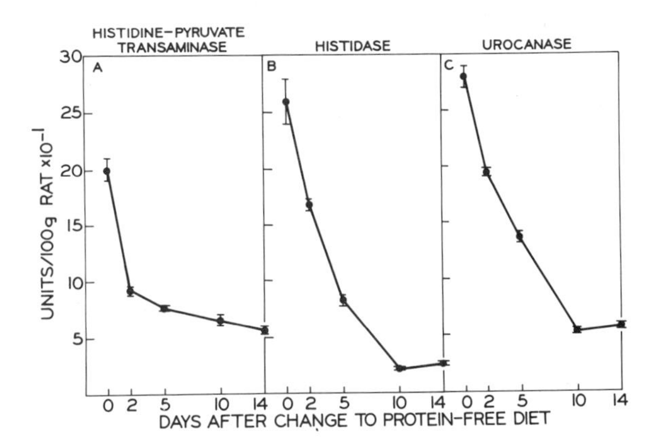

Figure 4-12. Changes in activities of histidine-degrading enzymes over time in livers of rats fed first an 80% casein diet and then a protein-free diet. (Schirmer and Harper, 1970)

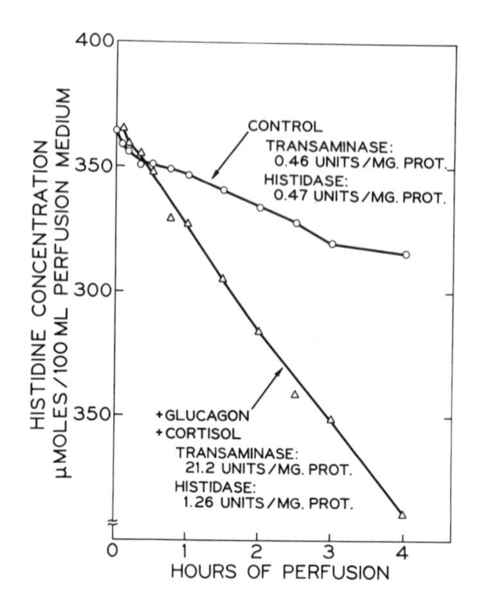

Figure 4-13. Removal of histidine from perfusion medium by livers from rats injected with glucagon and cortisol to induce high activities of histidine-degrading enzymes. (Morris and Harper, unpublished data)

kidney, the body can prevent the accumulation of amino acids in plasma and body fluids, if systems for removing excesses of amino acids are overloaded or defective, only through regulation of amino acid intake. Regulation of food intake, which is accomplished through the intact organism's functioning as a complex feedback system, is in reality regulation of transport. There is considerable evidence that elevated or depressed plasma amino acid concentrations lead to depressed food intake (Harper et al., 1970).

When the protein content of the diet of an animal is greatly increased an early response is depressed food intake. The depressed food intake is associated with elevations of plasma amino acid concentrations. Despite the depression in food intake, protein intake may still be elevated, and within a short time amino acid-degrading enzymes may increase in activity. After a period of adaptation plasma amino acid concentrations do not rise so high nor remain elevated so long and food intake gradually increases to close to its former level (Anderson et al., 1968). Thus, although the initial homeostatic response is depressed food intake, when protein intake remains elevated, amino acid-degrading enzymes respond.

The result is more rapid clearance of amino acids from the blood, and subsequent increase of food intake. Here is homeostasis par excellence.

When an animal is fed a low-protein diet containing an excessive amount of one amino acid, its food intake is severely depressed and metabolic adaptations occur slowly, if at all. Food intake may be depressed to the point where energy intake is inadequate. If the amount of amino acid in the diet is so high that the amount consumed still exceeds the capacity of the homeostatic mechanisms to remove it, blood amino acid concentration remains high, food intake continues to be depressed and the animal may become severely debilitated (Harper et al., 1970).

Although depressed food intake indicates that the capacity of homeostatic mechanisms has been exceeded, it is not necessarily an adverse effect: it is a response that contributes to the survival of animals fed a high-methionine diet. If the food intake of such animals is stimulated by lowering the environmental temperature, their intake of methionine is increased and they deteriorate more rapidly than those with depressed food intake in a warm environment (Beaton, 1967). Also, when animals are fed an unbalanced mixture of amino acids, their food intake is depressed but they usually survive well, adapt, and can eventually tolerate the diet. However, if the amino acid mixture is infused intravenously, so the amount of amino acids entering the body is not reduced by the depressed food intake, they do not adapt and the adverse effects are more severe (Peng and Harper, 1969).

Regulation of amino acid-degrading enzyme activity and regulation of food intake, the two major regulatory systems that contribute to the

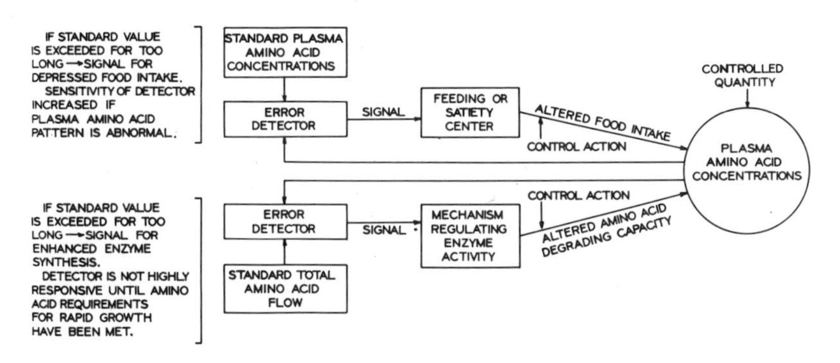

Figure 4-14. Schematic representation of feed-back loops for homeostatic regulation of plasma amino acid concentrations. (Harper, 1970)

maintenance of homeostasis in the organism faced with an amino acid load, can be portrayed diagrammatically as a double feedback loop (Figure 4-14). Changes in plasma amino acid concentrations are assumed to indicate the state of amino acid metabolism. If amino acids accumulate for too long, as detected by a food-intake regulating center, food intake is depressed to reduce the influx. Continued elevation of plasma amino acid concentrations also activates a mechanism which stimulates synthesis or depresses degradation of amino acid-degrading enzymes. Gradual adjustment of these mechanisms results in restoration of homeostasis with an altered level of amino acid metabolism.

Summary and Conclusions

In science we assume that all observers see the details identically. But when, as with control and regulation, the details for all observers are only fragmentary, it is valuable to have a basic framework into which the details can be fitted, much as an impressionist paints his canvas, and leaves it to his viewers to fill in the details. The concept of homeostasis provides such a framework in that the development of feedback diagrams serves to inventory the known and the unknown, serves to emphasize the need for quantification, and serves as a stimulus for research to reduce the number of unknowns. I should therefore like to conclude with Brobeck (1974) that the concept of homeostasis will hold its place in biology long after our individual capacities to maintain homeostasis have been exceeded.

References

Adibi, S. A., 1968. Influence of dietary deprivations on plasma concentration of free amino acids of man. *J. Appl. Physiol.* 25:52.

Anderson, H. L., Benevenga, N. J., and Harper, A. E., 1968. Associations among food and protein intake, serine dehydratase, and plasma amino acids. *Am. J. Physiol.* 214:1008.

Barcroft, J., 1932. La fixité du milieu intérieur est la condition de la vie libre. *Biol. Rev.* 7:24.

Beaton, J. R., 1967. Methionine toxicity in rats exposed to cold. *Can. J. Physiol. Pharm.* 45:329.

Benevenga, N. J., and Harper, A. E., 1970. Effects of glycine and serine on methionine metabolism in rats fed diets high in methionine. *J. Nutr.* 12:1205.

Bernard, C., 1878. *Leçons sur les phénomenes de la vie* (Vol. I and II). Librairie J.-B. Balliere et Fils, Paris.

Black, A. L., 1968. Modern techniques for studying the metabolism and utilization of nitrogenous compounds, especially amino acids. In *Isotope Studies on the Nitrogen Chain.* International Atomic Energy Agency, Vienna, p. 287.

Boctor, A. M., 1967. Some nutritional and biochemical aspects of tyrosine toxicity and lysine availability. Ph.D. diss., Massachusetts Institute of Technology.

Brobeck, J. R., 1974. Control, regulation, and homeostasis. In *The Control of Metabolism* (J. D. Sink, ed.). The Pennsylvania State University Press, University Park, p. 1.

Cannon, W. B. 1929. Organization for physiological homeostasis. *Physiol. Rev.* 9:399.

Coulson, R. A., and Hernandez, T., 1968. Amino acid catabolism in the intact rat. *Am. J. Physiol.* 215:741.

Elman, R., 1939. Time factor in retention of nitrogen after intraveneous injection of a mixture of amino acids. *Proc. Soc. Exptl. Biol. and Med.* 40:484.

Gaetani, S., Paolucci, A. M., Spadoni, M. A., and Tommassi, G., 1964. Activity of amino acid activating enzymes in tissues from protein-depleted rats. *J. Nutr.* 84:173.

Haider, M., and Tarver, H., 1969. Effect of diet on protein synthesis and nucleic acid levels in rat liver. *J. Nutr.* 99:443.

Harper, A. E., 1965. Effect of variations in protein intake on enzymes of amino acid metabolism. *Can. J. Biochem.* 43:1589.

Harper, A. E., 1968. Diet and plasma amino acids. *Am. J. Clin. Nutr.* 21:358.

Harper, A. E., 1970. Amino acid balance and food intake regulation. In *Parenteral Nutrition* (M. C. Meng and D. H. Law, eds.). Charles C Thomas, Springfield, Ill., p. 181.

Harper, A. E., and Benevenga, N. J., 1970. Effects of disproportionate amounts of amino acids. In *Proteins as Human Food* (R. A. Lawrie, ed.). The AVI Publ. Co., Westport, Conn., p. 417.

Harper, A. E., Benevenga, N. J., and Wohlhueter, R. M., 1970. Effects of disproportionate amounts of amino acids. *Physiol. Rev.* 50:428.

Henderson, L. J., 1928. *Blood—A Study in General Physiology.* Yale University Press, New Haven.

Jefferson, L. S., and Korner, A., 1969. Influence of amino acid supply on ribosomes and protein synthesis of perfused rat liver. *Biochem. J.* 111:703.

Jost, J.-P., Hsie, A., Hughes, S. D., and Ryan, L., 1970. Role of cyclic adenosine 3′,5-monophosphate in the induction of hepatic enzymes. *J. Biol. Chem.* 245:351.

Kaback, H. R., 1970. Transport. *Ann. Rev. Biochem.* 39:561.

Kim, J. H., and Miller, L. L., 1969. The functional significance of changes in activity of the enzymes, tryptophan pyrrolase and tyrosine transaminase, after induction in intact rats and in the isolated, perfused rat liver. *J. Biol. Chem.* 244:1410.

Knox, W. E., 1963. The adaptive control of tryptophan and tyrosine metabolism in animals. *Trans. N. Y. Acad. Sci.* 25:503.

Mallette, L. E., Exton, J. H., and Park, C. R., 1969. Effects of glucagon on amino acid transport and utilization in the perfused rat liver. *J. Biol. Chem.* 244:5724.

Mayer, J., and Vitale, J. J., 1957. Thermochemical efficiency of growth in rats. *Am. J. Physiol.* 189:39.

McLaughlan, J. M., and Illman, W. I., 1967. Use of free plasma amino acid levels for estimating amino acid requirements of the growing rat. *J. Nutr.* 93:21.

Miller, L. L., Burke, W. T., and Haft, D. E., 1956. Amino acid metabolism studies with the isolated perfused rat liver. In *Some Aspects of Amino Acid Supplementation* (W. H. Cole, ed.). Rutgers University Press, New Brunswick, p.44.

Munaver, S. M., and Harper, A. E., 1959. Amino acid balance and imbalance. II. Dietary level of protein and lysine requirement. *J. Nutr.* 69:58.

Munro, H. N., ed., 1964a. *The Role of the Gastrointestinal Tract in Protein Metabolism.* F. A. Davis Company, Philadelphia.

Munro, H. N., 1964b. General aspects of the regulation of protein metabolism by diet and by hormones. In *Mammalian Protein Metabolism* Vol. 1 (H. N. Munro and J. B. Allison, eds.). Academic Press, New York, p. 382.

Munro, H. N., 1968. Role of amino acid supply in regulating ribosome function. *Fed. Proc.* 27:1231.

Munro, H. N., 1970. Free amino acid pools and their role in regulation. In *Mammalian Protein Metabolism* (H. N. Munro, ed.). Academic Press, New York, p. 299.

Pearce, E. L., Sauberlich, H. E., and Baumann, C. A., 1947. Amino acids excreted by mice fed incomplete proteins. *J. Biol. Chem.* 168:271.

Peng, Y., and Harper, A. E., 1969. Amino acid balance and food intake: Effect of amino acid infusions on plasma and liver amino acids. *Am. J. Physiol.* 217:1441.

Peraino, C., Rogers, Q. R., Yoshida, M., Chen, M.-L., and Harper, A. E., 1959. Observations on protein digestion in vivo. II. Dietary factors affecting the rate of disappearance of casein from the gastrointestinal tract. *Can. J. Biochem. Physiol.* 37:1475.

Rogers, Q. R., and Harper, A. E., 1964. Digestion of proteins: Transfer rates along the gastrointestinal tract. In *The Role of the Gastrointestinal Tract in Protein Metabolism* (H. N. Munro, ed.). F. A. Davis Company, Philadelphia, p. 3.

Rogers, Q. R., and Harper, A. E., 1966. Protein digestion: Nutritional and metabolic considerations. In *World Review of Nutrition and Dietetics* (G. H. Bourne, ed.). Hafner, New York, p. 250.

Rosen, F., Roberts, N. R., and Nichol, C. A., 1959. Glucocorticosteroids and transimase activity. I. Increased activity of glutamic-pyruvic transimase in four conditions associated with gluconeogenesis. *J. Biol. Chem.* 234:476.

Sauberlich, H. E., 1961. Studies of the toxicity and antagonism of amino acids for weanling rats. *J. Nutr.* 75:61.

Sauberlich, H. E., and Baumann, C. A., 1946. The effect of dietary protein upon amino acid excretion by rats and mice. *J. Biol. Chem.* 166:417.

Scharrer, V. E., Erbersdobler, H., and Zucker, H., 1967. Untersuchungen uber den Verzehrsruckgang bei proteinreicher Ernahrung: Beziehungen zwischen Futterverzehr und freien Aminosauren im Plasma nach unterschiedlicher Adaptation an proteinreiche Futterung. *Z. Tierphysiol.* 22:265.

Schimke, R. T., 1962. Adaptive characteristics of urea cycle enzymes in the rat. *J. Biol. Chem.* 237:459.

Schimke, R. T., 1963. Studies on factors affecting the levels of urea cycle enzymes in rat liver. *J. Biol. Chem.* 238:1012.

Schimke, R. T., and Doyle, D., 1970. Control of enzyme levels in animal tissues. *Ann. Rev. Biochem.* 39:929.

Schimke, R. T., Sweeney, E. W., and Berlin, C. M., 1965. Role of synthesis and degradation in the control of rat liver tryptophan pyrrolase. *J. Biol. Chem.* 240:322.

Schirmer, M. D., and Harper, A. E., 1970. Adaptive responses of mammalian histidine-degrading enzymes. *J. Biol. Chem.* 245:1204.

Sidransky, H., Sarma, D. S. R., Bongiorno, M., and Verney, E., 1968. Effect of dietary tryptophan on hepatic polyribosomes and protein synthesis in fasted mice. *J. Biol. Chem.* 243:1123.

Soliman, A.-G., and Harper, A. E., 1969. Contribution of tissue protein degradation to amino acid pools in rat tissues. *Fed. Proc.* 28:302.

Spolter, P. D., and Harper, A. E., 1961. Utilization of injected and orally administered amino acids by the rat. *Proc. Soc. Exptl. Biol. and Med.* 106:184.

Tallan, H. H., Moore, S., and Stein, W. H., 1954. Studies on the free amino acids and related compounds in the tissues of the cat. *J. Biol. Chem.* 211:927.

Tannous, R. I., 1963. Metabolic studies on leucine, isoleucine and valine antagonism in the rat. Sc.D. diss., Massachusetts Institute of Technology.

Tews, J. K., and Harper, A. E., 1969. Transport of nonmetalizable amino acids in rat liver slices. *Biochim. Biophys. Acta* 183:601.

Tews, J. K., Woodcock, N. A., and Harper, A. E., 1970. Stimulation of amino acid transport in rat liver slices by epinephrine, glucagon and adenoisine 3′,5′-monophosphate. *J. Biol. Chem.* 245:3026.

Waterlow, J. C., 1968. Observations on the mechanism of adaptation to low protein intakes. *The Lancet* 2:1091.

Wohlhueter, R. M., and Harper, A. E., 1970. Coinduction of rat liver branched chain α-keto acid dehydrogenase activities. *J. Biol. Chem.* 245:2391.

Wool, I. G., Stirewalt, W. S., Kurhara, K., Low, B., Bailey, P., and Oyer, D., 1968. Mode of action of insulin in the regulation of protein biosynthesis in muscle. In *Recent Progress in Hormone Research* (E. B. Astwood, ed.). Academic Press, New York, p. 139.

5

Regulatory Function of Biological Membranes

Stuart Patton

Membranes are the structural-functional aspect of the cell that carry out its various life processes. Membranes represent the principal immediate effects of the genetic message which lead to the infinite variety of cell types comprising all of life. When one speaks of regulatory functions of membranes one is concerned with the bedrock of regulation; for the synthesis, transport, and utilization of any metabolite ultimately relates to membranes. For mastery of life at its simplest level, the cell, an understanding of membranes seems no less important than a knowledge of the gene. And it is evident that many workers are in hot pursuit of this goal. The number of published papers and journals on the subject of membranes is steadily increasing. In 1967, *Biochimica et Biophysica Acta* started a Biomembranes section; the number of pages in that section had quadrupled by 1972.

Membrane Structure and Function

It is also evident that we do not yet know the molecular architecture of membranes. Even now the subject is strongly debated. (For a popular presentation see *Chemical and Engineering News,* 25 May 1970.) The crux of the argument concerns the arrangement of the lipid and protein molecules making up the membranes. One school of thought, originating with the Danielli-Davson hypothesis (1935), proposes that the fundamental structure in all membranes is a bimolecular layer of lipid molecules to which various proteins are bound. A second theory of the structure, based on models and evidence by Benson (1966) and Green et al.

(1967), holds that the basic matrix is an aggregate of protein subunits into which lipid molecules are fitted and bound in various ways (Figure 5-1). There are proponents who feel that both such structures exist in living systems and conceivably within the same type of membrane. For pertinent reviews, see Stoeckenius and Engelman (1969), Korn (1969), and O'Brien (1967).

While I do not have anything decisive to add to the argument, a few points about membrane structure are worth pondering. It would be rather remarkable if Nature developed only one rather rigid solution to the problem of membrane structure, which after all is the basis of life in the cell. A single type of structural arrangement regarding lipid and protein seems especially unlikely when the amount of lipid in a membrane may range from essentially zero for the gas vacuole membrane of blue-green algae (Jones and Jost, 1970) to 80% for myelin (O'Brien, 1967). Coincidentally, we have noted that the swim bladder (outer wall) of the sheepshead fish *(Pimelometopon pulcher)* is a thick, extremely tough, proteinaceous membrane which contains no lipid. Such findings would suggest that membranes which act for retention of gases do not contain lipids. The corollary, that gas transport requires lipid-containing membranes, might also hold. It would be impossible to invoke a lipid bilayer structure in a membrane with little or no lipid, and it would be almost impossible to avoid such a structure, at least in some degree, in a membrane which is 80% lipid on a dry basis.

A more significant issue may lie in the increasing evidence of membrane-bound enzymes that are strongly associated with and/or are activated by lipids. There is the classic evidence that mitochondrial enzymes can be reactivated by restoring extracted lipids (Fleischer et al., 1967a). In addition the phospholipid, cardiolipin, is specifically bound to cytochrome oxidase and required for its activity and structure in the mitochondrial membrane (Awasthi et al., 1970). Moreover, 5′-nucleotidase, a marker enzyme for the plasma membrane, has been recovered as a lipoprotein in which sphingomyelin makes up over 90% of the lipid (Widnell and Unkeless, 1968). Another example: xanthine oxidase has been shown to lose activity when rendered lipid-free and to gain activity when treated with lipids (Mitidieri et al., 1970). One is taxed to explain these rather specific effects of lipids on membrane-bound enzymes on the basis of a uniformly conceived structure such as a bilayer. Invoking the potential interaction of the whole lipid molecule in explanations of membrane function, rather than just the polar head groups, has the advantage that many more variations in (reactive) tertiary structure of membrane proteins could be induced through lipid-protein interaction. Here the suggestion is that a limited number of proteins might involve a wide range of enzymatic activities.

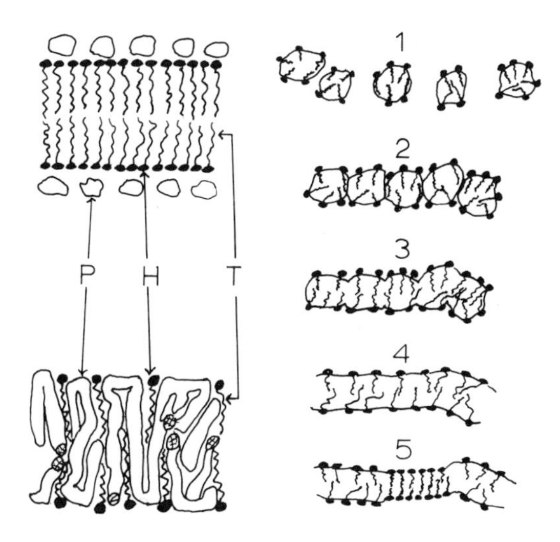

Figure 5-1. Membrane models showing location of protein (P) and lipid molecules with polar head groups (H) and hydrocarbon tails (T). Upper left, the Danielli-Davson (1935) bilayer model; lower left, the Benson (1966) model with protein matrix and interdigitated lipids; right a conception of membrane formation from lipoprotein subunits. At stage 5, membrane shows evolution of areas with features of both models.

If there is some uncertainty about membrane structure, there are even more questions to raise about function. For example, membrane transport accounts for selective movement of substances ranging from small ions to large macromolecules, from highly polar to highly nonpolar materials, into and out of the cell without loss of cell integrity. There is only speculation thus far as to how this happens at the molecular level. In the case of small molecules such as amino acids and sugars, it is known that there are specific carrier proteins in cell membranes which selectively combine with the metabolite molecules and are required for their transport across the membrane. The selectivity of the transport proteins, when coupled with the specificity of acceptor enzymes for the metabolites within the cell, provides a highly efficient screening of exogenous compounds (Berlin, 1970). One particularly interesting transport enigma concerns the passage of immune globulins from bovine mother to young via milk. These large molecules must be passed across many types of cell membranes, from blood to milk and from ingested milk back to blood.

My point in belaboring the inadequacy of our knowledge of membranes is to make clear how incautious it may be to talk about their regulatory functions. It is obvious that membranes regulate all manner of things, including you and me, but how they do this will keep some of us busy a long while. Some of the proposed and established general functions of specific cell membranes are shown below.

Nuclear membrane
- Retention of nuclear material
- Regulation of transport to and from nucleus
- Giving rise to other cell membranes

Mitochondrial membrane
- Cell respiration
- Oxydative phosphorylation
- Metabolite synthesis and transformation

Endoplasmic reticulum membrane
- Synthesis of lipids, proteins, enzymes, etc.
- Serving as source of other cell membranes and membrane components

Golgi membrane
- Combining and packing of products of the endoplasmic reticulum into secretory vesicles
- Transformation of membranes
- Glycosylation of lipids, proteins, and carbohydrates

Plasma membrane
- Retention of cell contents
- Regulation of transport into and out of the cell
- Recognition of and communications with other cells
- Regulation of secretory processes

Much is known about the function of the mitochondrion (Ernster and Drahota, 1969), and knowledge of the relation between membrane structure and function has been pushed further with this organelle than with any other membrane of the cell. In addition, regulatory functions of membranes is the subject of a book (Jarnefelt, 1968) which makes a much more extensive contribution to the subject than is possible here. However, I would like to point out a few properties of some membranes which relate to their structure and function. Out of the vast number of different cell types, it will be necessary to exemplify and generalize in so far as possible. For this purpose, I draw of the hepatocyte and the lactating mammary cell. As might be expected, more research has been conducted on the membranes of rat liver than any other animal tissue. As a consequence, the membrane preparations from the rat hepatocyte are the purest and most completely characterized. While data on membranes of the lactating mammary cell are not adequate, our extensive knowledge of milk and its secretion enable important insights into membrane function.

Cholesterol and Sphingomyelin in the Cell Membrane System

Membranes which are highly structural, such as myelin or the envelope of the erythrocyte, are high in cholesterol, sphingolipids, and total lipids. For example, myelin is 80% lipid on a dry basis. It contains roughly 40 mole per cent cholesterol and 26 mole per cent sphingolipid (O'Brien, 1967). Its primary function appears to be serving as an electrical insulator for the nerve axon. On the other hand, highly functional membranes such as those of the mitochondrion or endoplasmic reticulum contain much lower levels of these lipids. In this connection, it was of interest to study the membranes of a single cell type. The cholesterol-sphingomyelin relationship for the rat hepatocyte membranes are shown in Figure 5-2. It is evident from these data that there is a positive

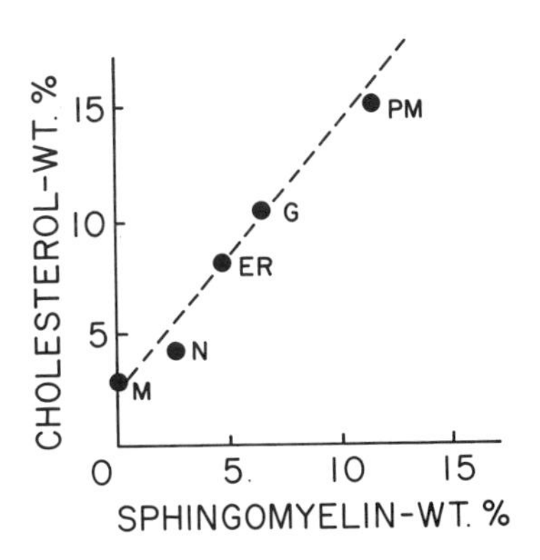

Figure 5-2. The relationship (weight per cent) between cholesterol and sphingomyelin in membranes of the rat hepatocyte. M *mitochondrion;* N *nucleus;* ER *endoplasmic* G PM *plasma membrane (Patton, 1970).*

Table 5-1. Amounts of Cholesterol, Spingomyelin, and Total Lipids in Membrane of Rat Liver

Membrane	*Cholesterol*		*Spingomyelin*		*Total lipid (dry basis)*	
	Wt. %	*Source*	*Wt. %*	*Source*	*Wt. %*	*Source*
Mitochondria	3.0	Schwartz et al., 1961	Nil	Schwartz et al., 1961	15-24*	Dod and Gray, 1968; Fleischer et al., 1967b
Nuclear	4.3	Franke et al., 1970 Kleinig, 1970	2.7	Franke et al., 1970 Kleinig, 1970	19	Franke et al., 1970 Kleinig, 1970
Endoplasmic reticulum	8.2	Glaumann and Dallner, 1968	4.9	Glaumann and Dallner, 1968	24-29	Glaumann and Dallner, 1968; Yunghans et al., 1970
Golgi	10.3	Keenan and Morre, 1970	6.6	Keenan and Morre, 1970	40†	Yunghans et al., 1970
Plasma membrane	18.9	Keenan and Morre, 1970	9.8	Keenan and Morre, 1970	40	Ray et al., 1969

*Including data for bovine liver, heart, and kidney.
†Including data for lipids from contaminating serum ß-lipoproteins.

correlation between the presence of cholesterol and that of sphingomyelin in the various cell membranes. Interestingly, the maximum levels of both are found in the plasma membrane, which has the important structural function of holding the cell together. In general the total lipid contents (dry basis) of the membranes increase as the cholesterol and sphingomyelin increase (Table 5-1).

Two ways of viewing the observed cholesterol-sphingomyelin relationship are: (1) they form a stable complex in membranes (possibly involving synthesis of one or the other) and thus tend to vary correlatively, and (2) there is an evolution or flow of membranes within the cells in which there is a net loss of other components in the membranes and a net increase in cholesterol, sphingomyelin, and total lipid. If this latter hypothesis is true the data predict that the membrane transformation is: nuclear → endoplasmic reticulum → Golgi → plasma membrane. There already exists some support for this scheme. It is known that smooth endoplasmic reticulum originates from rough (ribosomal-laden) (Dallner et al., 1967), that the Golgi apparatus gives rise to secretory vesicles the membranes of which merge with the plasma membrane (Bargmann and Knoop, 1959) (Figure 5-3) and that the Golgi apparatus appears to serve the function of transforming membranes of the endoplasmic reticulum to those suitable for secretory vesicles (Grove et al., 1968; Keenan and Morre, 1970). We have presented evidence of membrane flow for the lactating mammary cell (Patton and Hood, 1969). As an active secretory cell (serum lipoproteins), the hepatocyte may also involve a program of membrane conversion.

From a biomedical standpoint, it is interesting that cellular events may tend toward the accumulation of cholesterol, sphingomyelin, and total lipids in the plasma membrane. All three of these parameters are elevated in atheromatous plaques (Buck and Rossiter, 1952), and the plasma membrane has been suggested as a possible origin of the accumulating sphingomyelin (Eisenberg et al., 1969; Portman and Alexander, 1970).

Mitochondrial membranes represent a special case in that they acquire much of their lipids by transfer from endoplasmic reticulum (Wirtz and Zilversmit, 1968; Jungalwala and Dawson, 1970). Cardiolipin appears to be the only phospholipid which is synthesized by the mitochondrion. Endoplasmic reticulum in the rat hepatocyte is known to be the principal site of cholesterol and phospholipid syntheses (Chesterton, 1968; Jungalwala and Dawson, 1970). Our research of the literature indicates that mitochondria contain very little of either cholesterol or sphingomyelin, and what little found may be there either by exchange or by contamination.

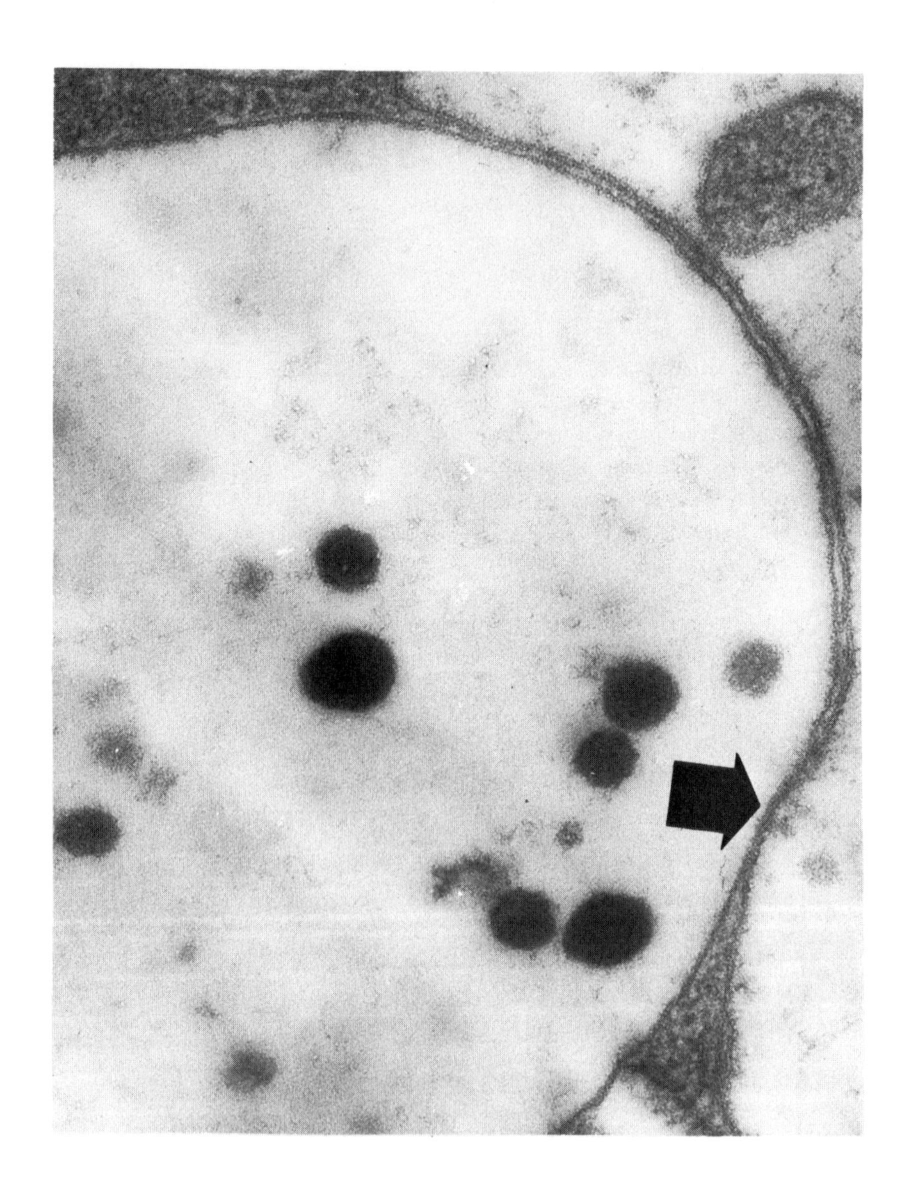

Figure 5-3. Electron photomicrograph of a section through lactating bovine mammary tissue, showing a Golgi vesicle containing (dark) granules of casein. This vesicle is close to the instant of secretion. Its limiting membrane is pressing against the plasma membrane of the cell (arrow). (× *150,000; courtesy of K. E. Beery)*

The Plasma Membrane of the Lactating Mammary Cell

The lactating mammary cell provides an interesting object of study from the standpoint of extracellular environment in relation to plasma membrane function. (For a current understanding of how the lactating mammary cell functions see *Scientific American* 221:58, July, 1969.) The cell's habitat is the alveolus (Figure 5-4). At its base, the cell is oriented toward the capillaries which are supplying it with nutrients and substrates for milk synthesis. Normally, this aspect of the plasma membrane is devised for transport of a unique group of materials into the cell. At the apex of the cell, the plasma membrane is regulating the secretion process or transport of materials out of the cell. This may concern no more than two basic mechanisms: i.e., fat droplet secretion by envelopment with plasma membrane followed by a pinching-off, and the emptying through the membrane of secretory vesicles containing the dissolved and dispersed nonfat phase of milk. Adjacent to the cell are other like cells so that the plasma membrane of one cell abuts that of another

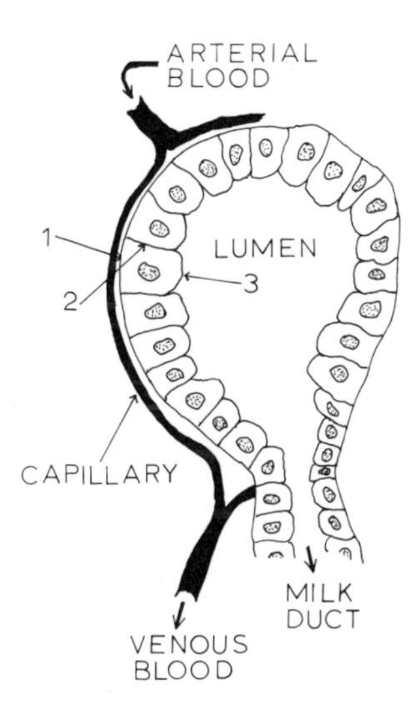

Figure 5-4. Scheme of the alveolus, showing shell of lactating mammary cells around lumen into which milk is secreted. Flow of milk is from lumen into duct system (arrow). *Differentiated environment of mammary cell plasma membrane is indicated at 1, where membrane takes up metabolites from blood, 2, where adjacent cells communicate, and 3, where secretion takes place.*

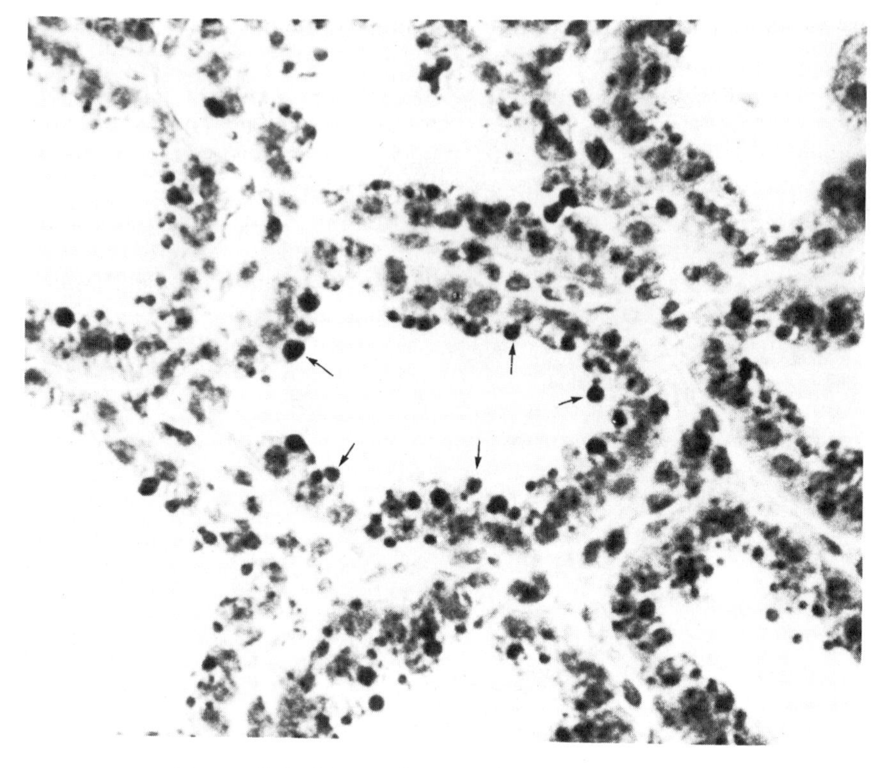

Figure 5-5. Photomicrograph of a section through lactating bovine mammary tissue, showing an alveolus with all cells lactating and containing milk fat droplets (arrow) *at about the same size and (apical) position in all of the cells.*

and transport across these membranes represents communication between cells. The implication is that plasma membrane has quite different orientations and functions at various locations in the cell. Moreover, the secretion process should dictate a rapid turnover of plasma membrane in the apex of the cell, whereas a more stable situation may exist at the base and sides of the cell. During the "drying-up" period or involution of the mammary gland, the function of these membranes must change markedly in that the movement of material is reversed from the alveolar lumen and cell back into the blood. We do not know at this time how the plasma membrane achieves these special functional capabilities on the various areas of the cell surface. With the exception of involution, which induces microvilli formation in the lumen area, we do not know what differences in membrane structure and composition are involved at the various areas.

Membranes and Communications between Cells

It is very difficult to divorce a consideration of regulation by membranes from regulation by hormone or gene actions. Of course, ribosomes, which are the product of gene action, represent a most vital part of the endoplasmic reticulum, so that gene action and product synthesis by membranes are integrated regulatory functions. However, in the case of hormone action the precise mechanism is not known and it is reasonable to assume that peptide and corticoid hormones may have quite different molecular modes of action. If the site of action of peptide-type hormones is the plasma membrane, as some suggest, the regulatory role of this membrane in the functioning of the cell is profound indeed.

Observations by Saacke and Heald (1969), confirmed by others (Stewart, 1970; Keenan et al., 1970), indicate unique control mechanisms for the lactating mammary cell at the level of the alveolus. Cells of a given alveolus are seen to be synchronized—i.e., either all cells are lactating or all are dormant. A further point of interest is that all cells of an alveolus will show synchrony in their phase of fat globule synthesis and secretion (Figure 5-5). Even though the lactating cell is a result of consort hormonal action, it is remarkable that neither a mixture of active and inactive cells nor a mixture of cells having different phases and levels of activity are obtained within any particular alveolus. This suggests a type of communication among cells in an alveolus and renders the site and mechanism of prolactin action a particularly intriguing question. Prolactin is the culminating hormone which brings the dormant but prepared mammary epithelial cell into lactation (Mills and Topper, 1970; Keenan et al., 1970).

Gaps in the plasma membranes between adjacent cells, and cells containing two nuclei next to ones with no nuclei, have been observed in lactating alveoli (R. G. Saacke and C. W. Heald, personal communication). The suggestion is that small cytoplasmic continuities (pores) between cells may be a factor in cell synchrony. In this case the tendency of a membrane to form pores would constitute an important regulatory mechanism.

Summary and Conclusions

Membrane science is clearly a vast and challenging field and our knowledge of it is meager. In essence, animal science *is* membrane science. The steady growth of the field of membrane research is heartening, but if we are to gain a better understanding of animal metabolism—and one can equate that with bettering meat production, milk production, processing properties of foods, animal health, and man's well-being, as well as assuring his survival—even greater emphasis on membrane research is needed.

This work was supported in part by Grant HE 03632, from the U. S. Public Health Service. Thanks are due A. A. Benson for discussion of membrane structure and function, and T. W. Keenan and R. G. Saacke for ideas on the biochemistry and ultra structure of the lactating mammary cell.

References

Awasthi, Y. C., Chuang, T. F., Keenan, T. W., and Crane, F. L., 1970. Association of cardiolipin and cytochrome oxidase. *Biochem. Biophys. Res. Comm.* 39:822.

Bargmann, W., and Knoop, A., 1959. Uber die Morphologie der Milchsekretion. I. Licht und elektronenmikroskopische Studien an der Milchdruse der Ratte. *Z. Zellforsch. Mikrosk. Anat.* 49:334.

Benson, A. A., 1966. On the orientation of lipids in chloroplasts and cell membranes. *J. Am. Oil Chemists Soc.* 43:265.

Berlin, R. D., 1970. Specificities of transport systems and enzymes. *Science* 168:1539.

Buck, R. C., and Rossiter, R. J., 1952. Lipids of normal and atherosclerotic aortas. *Arch. Path.* 51:224.

Chesterton, C. J., 1968. Distribution of cholesterol precursors and other lipids among rat liver intracellular structures. *J. Biol. Chem.* 243:1147.

Dallner, G., Siekovitz, P., and Palade, G. E., 1966. Biogenesis of endoplasmic reticulum membranes. I. Structural and chemical differentiation in developing rat hepatocyte. *J. Cell Biol.* 30:73.

Danielli, J. F., and Davson, H., 1935. A contribution to the theory of permeability of thin films. *J. Cell Comp. Physiol.* 5:495.

Dod, B. J., and Gray, G. M., 1968. The lipid composition of rat liver plasma membrane. *Biochim. Biophys. Acta* 150:397.

Eisenberg, S., Stein, Y., and Stein, O., 1969. Phospholipases in arterial tissue. III. Phosphatide acyl-hydrolase and sphingomyelin choline phosphohydrolase in rat and rabbit aorta in different age groups. *Biochim. Biophys. Acta* 176:557.

Ernster, L., and Drahota, Z., 1969. *Mitochondrial Structure and Function.* Academic Press, New York.

Fleischer, S., Fleischer, B., and Stoeckenius, W., 1967a. Fine structure of lipid-depleted mitochondria. *J. Cell Biol.* 32:193.

Fleischer, S., Rouser, G., Fleischer, B., Casu, A., and Kritchevsky, G., 1967b. Lipid composition of mitochondria from bovine heart, liver and kidney. *J. Lipid Res.* 8:170.

Franke, W. W., Deumling, B., Ermen, B., Jarasch, E.-D., and Kleinig, H., 1970. Nuclear membranes from mammalian liver. I. Isolation procedure and general characterization. *J. Cell Biol.* 46:379.

Glaumann, H., and Dallner, G., 1968. Lipid composition and turnover of rough and smooth microsomal membranes in rat liver. *J. Lipid Res.* 9:720.

Green, D. E., Allman, D. W., Bachmann, E., Baum, H., Kopaczyk, K., Korman, E. F., Lipton, S., Mac Lennan, D. H., Mc Connell, D. G., Perdue, J. F., Rieske, J. S., and Tzagoloff, A., 1967. Formation of membranes by repeating units. *Arch. Biochem. Biophys.* 119:312.

Grove, S. N., Bracker, C. E., and Morre, D. J., 1968. Cytomembrane differentiation in the endoplasmic reticulum-Golgi apparatus vesicle complex. *Science* 161:171.

Jarnefelt, J., 1968. *Regulatory Function of Biological Membranes.* Elsevier, Amsterdam.

Jones, D. D., and Jost, M., 1970. Isolation and chemical characterization of gas vacuole membranes from *Microcystis aeruginosa* Kuetz. emend Eleukin. *Arch. Mikrobiol.* 70:43.

Jungalwala, F. B., and Dawson, R. M. C., 1970. Phospholipid synthesis and exchange in isolated liver cells. *Biochem. J.* 117:481.

Keenan, T. W., and Morre, D. J., 1970. Phospholipid class and fatty acid composition of Golgi apparatus isolated from rat liver and comparison with other cell fractions. *Biochem.* 9:19.

Keenan, T. W., Saacke, R. G., and Patton, S., 1970. Prolactin, the Golgi apparatus and milk secretion: a brief interpretive review. *J. Dairy Sci.* 53:1349.

Kleinig, H., 1970. Nuclear membranes from mammalian liver. II. Lipid composition. *J. Cell Biol.* 46:396.

Korn, E. D., 1969. Cell membranes: structure and synthesis. *Ann. Rev. Biochem.* 38:263.

Mitidieri, E., Affonso, O. R., and Ribeiro, L. P., 1970. Role of phospholipids in xanthine dehydrogenase activity. *Enzymologia* 38:161.

Mills, E. S., and Topper, Y. J., 1970. Some ultrastructural effects of insulin, hydrocortisone and prolactin on mammary gland explants. *J. Cell Biol.* 44:310.

O'Brien, J. S., 1967. Cell membranes—composition; structure: function. *J. Theoret. Biol.* 15:307.

Patton, S., 1970. The correlative relationship of cholesterol and sphingomyelin in cell membranes. *J. Theoret. Biol.* 29:489.

Patton, S., and Hood, L. F., 1969. Biogenesis of milk lipids. In *Lactogenesis: The Initiation of Milk Secretion at Parturition* (M. Reynolds and S. J. Folley, eds.). University of Pennsylvania Press, Philadelphia, p. 121.

Portman, O. W., and Alexander, M., 1970. Metabolism of sphingolipids by normal and atheroscleratic aorta of squirrel monkeys. *J. Lipid Res.* 11:23.

Ray, T. K., Skipski, V. P., Barclay, M., Essner, E., and Archibald, F. M., 1969. Lipid composition of rat liver plasma membranes. *J. Biol. Chem.* 244:5528.

Saacke, R. G., and Heald, C. W., 1969. Histology of lipid synthesis and secretion in the bovine mammary gland. *J. Dairy Sci.* 52:917.

Schwartz, H. P., Driesbach, L., Barrionuevo, M., Kleschick, A., and Kostyke, I., 1961. The effect of ionizing irradiation on the lipid composition of the liver mitochondria of rats. *Arch. Biochem. Biophys.* 92:133.

Stewart, P. S., 1970. A study of the formation and structure of the milk fat globule membrane. Master's thesis, University of Guelph, Ontario, Canada.

Stoeckenius, W., and Engelman, D. M., 1969. Current models for the structure of biological membranes. *J. Cell Biol.* 42:613.

Widnell, C. C., and Unkeless, J. C., 1968. Partial purification of a lipoprotein with 5′ nucleotidase activity from membranes of rat liver cells. *Proc. Nat. Acad. Sci.* 61:1050.

Wirtz, K. W. A., and Zilversmit, D. B., 1968. Exchange of phospholipids between liver mitochondria and microsomes *in vitro*. *J. Biol. Chem.* 243:3596.

Yunghans, W. N., Keenan, T. W., and Morre, D. J., 1970. The composition of Golgi apparatus and comparisons with other cell fractions from rat liver. *Exp. Mol. Path.* 12:36.

6

Food Intake, Energy Balance, and Homeostasis

B. R. Baumgardt

The concept of "homeostasis" can be applied to the regulation of energy balance, as most adult animals tend to maintain a stable body weight over time in spite of changes in activity and environment. Since the source of energy for maintenance of energy balance can be related to the energy in food, it follows that the control of food intake can also be considered in terms of homeostasis. Yamamoto and Brobeck (1965) have pioneered in the application of feedback analysis to whole-animal systems, a tool we have found very useful for coordinating existing knowledge and as an indicator of needed research.

A simplified chart will be used to present my impressions of the state of knowledge in this area. I have been using this approach for some time as have others (Baile, 1968; 1971).

The following topics will be considered: (1) evidence for regulation of energy balance, in particular, body fat stores; (2) involvement of the central nervous system, in particular, the hypothalamus; (3) the nature of the feedbacks involved in the short-term and long-term control of food intake; and (4) varying levels of regulation, with comments about some errors in regulation of energy balance. I will not attempt a comprehensive review of literature since several recent reviews are available. Key points will be documented with examples from the available literature, with emphasis on recent publications. In other areas some speculation is needed.

An overview of some of the types of factors involved in the regulation of energy balance is shown in Figure 6-1. Energy balance is the component regulated. Physiological and metabolic events during and following a meal result in certain meal or short-term feedbacks which are detected and transduced. Detector signals are balanced against a reference input signal. The net feedback from this comparator will determine whether an actuating signal will be generated. Such an actuating signal would be received by

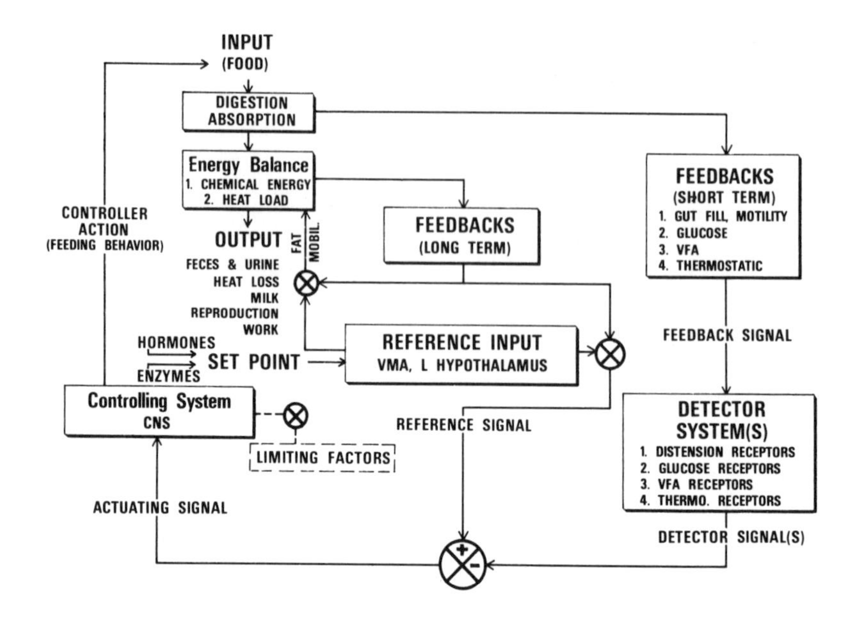

Figure 6-1. Schematic representation of interrelationships in the control of food intake and the regulation of energy balance.

a controlling system or center that serves to integrate all inputs that influence feeding behavior. The ultimate decision regarding feeding behavior (i. e., to eat or not to eat) is made here. Certain other feedbacks, arising from the metabolic storehouses within the animal, directly or indirectly influence the reference signal (long-term regulation).

Regulation of Energy Balance

Energy balance, generally considered the controlled component, should be examined first. Animals, in a given physiological state, adjust the amount of food consumed in relation to the energy content of the diet and thus appear to eat for calories (Baumgardt, 1970). We have been attempting to quantitate the relationship between caloric content of the diet and food and energy intake. Data from a study with young rats (Peterson and Baumgardt, 1971a) are presented in Figure 6-2. The younger rat, 32 days of age, has the higher energy requirement per unit of metabolic weight. However, at both 33 and 52 days of age (means) the rats

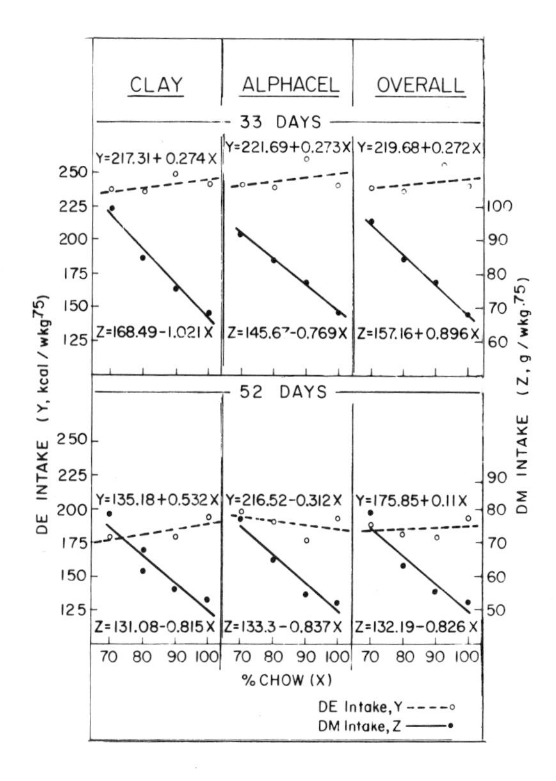

Figure 6-2. Digestible energy (DE) intake (Y) and dry matter (DM) intake (Z) of rats at 33 and 52 days of age fed laboratory chow diets diluted with 0, 10, 20, and 30% kaolin (clay) and alphacel.

compensated for up through 30% dilution of their diets with either clay or alphacel by increased food intake. The regression coefficients for DE (digestible energy) intake did not differ from zero. In a subsequent experiment, a larger range of dilutions—0, 20, 30, 35, 40, 45, 50, and 55%—was imposed. Net body-weight gains were maintained through the 40% dilution with kaolin (most dense diluent), and the 30% dilution with alphacel (intermediate density), but only the 20% dilution with perlite (least dense diluent).

In another study (Peterson and Baumgardt, 1971b), diet dilution (0 to 70% perlite) was imposed on rats at three physiological stages (Figure 6-3). Lactating rats nursing 10 pups per litter were unable to maintain energy intake with even the 10% dilution. Weanling and mature rats were able to maintain DE intake through the 45% dilution; this was

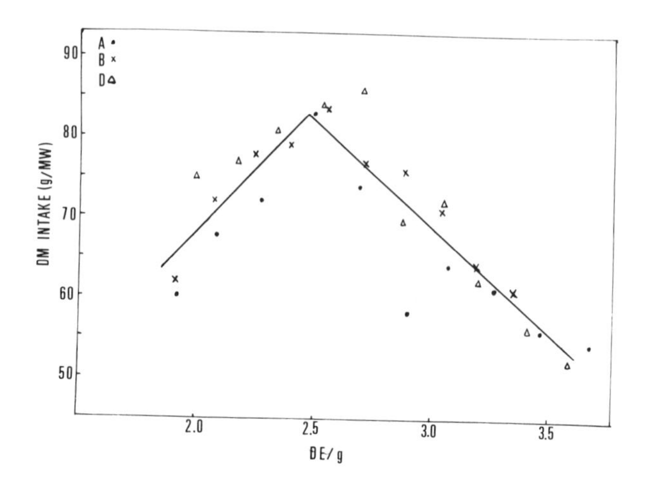

Figure 6-3. Digestible energy (DE) intake (Kcal/day $W_{kg}^{0.75}$) by mature, weanling, and lactating rats fed diets diluted 0, 10, 20, 30, 40, 50, 60, and 70% with perlite.

equivalent to a DE value of about 1.8 Kcal/g or 2.6 Kcal/ml (based on water-displacement density). These studies demonstrate a marked homeostatic ability of the rat to maintain energy intake. At the same time they show that this ability to adjust can be overcome by extensive diet dilution where, presumably, fill or distension in some part of the gut becomes the first limiting factor. Caloric density (kilocalories per milliliter) was a better predictor of DE intake and net body-weight gain than DE per unit weight of diet (r = 0.95, as against 0.86) (Peterson and Baumgardt, 1971a). An explanation is that gastrointestinal capacity becomes a factor limiting intake in relation to both the energy content of the diet (kilocalories per gram) and the space occupied in the gut by that diet (density). The minimum caloric density at which these growing rats could maintain DE intake was between 2.66 and 2.89 Kcal/ml.

We have made comparative studies with ruminants (Dinius and Baumgardt, 1970). Figure 6-4 shows the DM (dry matter) intake by sheep of diets diluted from 5 to 50% with undigestible diluents. DM intake increases up to 35% dilution. Figure 6-5 shows that on the same

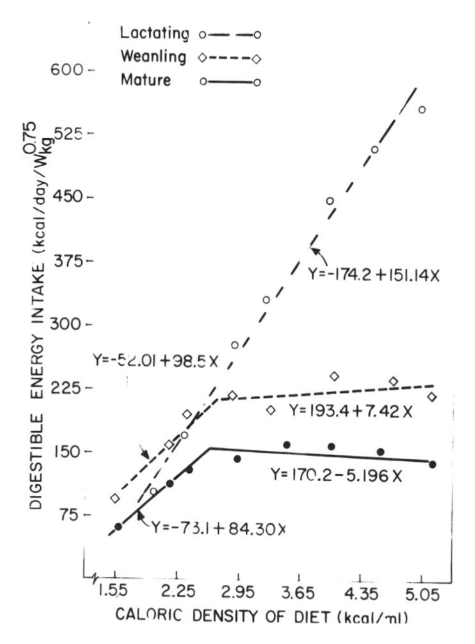

Figure 6-4. Feed dry matter (DM) intake (g/MW) of sheep fed a basal concentrate mixture diluted from 5 to 50% at 5% increments with each of three diluents: (A) oak sawdust, (B) oak sawdust with constant 3% kaolin clay, (D) oak sawdust with nitrogen kept constant at 17.4% crude protein. (Dinius and Baumgardt, 1970)

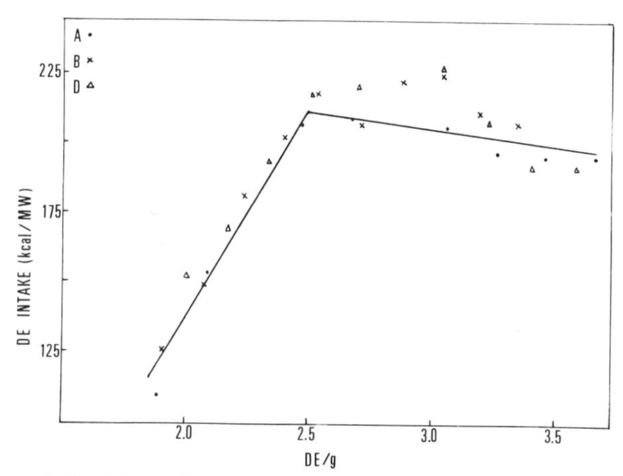

Figure 6-5. Digestible energy (DE) intake of sheep fed rations varying in energy content as described for Figure 6-4. (Dinius and Baumgardt, 1970)

diets, energy intake was also maintained up to the 35% dilution. This dilution represents DE values of 2.5 Kcal/g or 3.5 Kcal/ml.

Very unpalatable foods can prevent the maintenance of energy homeostasis. However, in another study (Welton and Baumgardt, 1970), we have shown that only about 6% of the "expected difference" in intake between the 30% and the 50% sawdust-diluted diets was due to palatability.

Clearly, animals as diverse as rats and sheep tend to eat for calories. Available evidence suggests that body fat stores is the parameter being regulated, a subject that will be discussed later in this chapter.

Hypothalamus and Food Intake

The classic work of Hetherington and Ranson (1939) opened the way for numerous studies which show the important role of the hypothalamus in controlling food intake (see "Reference Input" and "Controlling System," Figure 6-1).

The general visualization of the role of hypothalamic centers is that satiety signals are received by the ventromedial area (VMA), the satiety center, and that the lateral area (LH) remains active unless inhibited by signals from an activated VMA (see Figure 10-3 on page 182 in Tepperman, 1968). In the hunger state there is no satiety signal, no inhibition of the feeding center; thus, it remains active and the animal eats. After a meal there is a satiety signal which inhibits the feeding center; thus, the animal does not eat. Lesioning or stimulation of the appropriate areas provides interruption of signals or false signals with concomitant "errors" in feeding behavior and energy intake.

Electrolytic lesions in the VMA result in hyperphagia, and lesions in the LH or feeding area result (at least temporarily) in aphagia (DeGroot, 1967). These effects have been noted in a number of laboratory animal species (DeGroot, 1967) and in goats (Baile et al., 1967). Reports of hypothalamic hyperphagia in domestic livestock which have a propensity for fattening are fewer in number, perhaps because of fewer experimental attempts. Temporary hyperphagia has been produced in rats, cats, and goats (Baile and Mayer, 1966) by infusion or perfusion of the ventriculocisternal system with a depressant. California workers (Khalaf et al., 1970) have induced temporary hyperphagia in swine by injection of pentobarbitol into the VMA of already sated animals. We have similar results in sheep (Baumgardt and Peterson, 1970) and calves (unpublished data); an example of the sheep data is shown in Figure 6-6.

During the session interval on no-treatment days, mean food intake was 75 g, whereas on infusion days the mean intake was increased to 612 g for the same time interval. These results suggest a pliability of hypothalamic control in farm animal species similar to that observed in laboratory animals. Convincing demonstrations of extended hyperphagia in farm livestock due to VMA lesions are still lacking, although Khalaf (1969) has shown a moderate response in swine. Hypothalamic involvement will be discussed further below, especially in terms of the set-point concept.

Short-Term (Meal) Control of Food Intake

Feedback signals are generated by the consumption of a meal and the subsequent digestion, passage, and metabolism (Figure 6-1). In terms of mode of action, control systems can be described as either short-term or long-term. Short-term control is concerned with the initiation and cessation of individual meals. It is apparent that errors in adjustment of energy intake to energy expenditure are made in individual meals, on

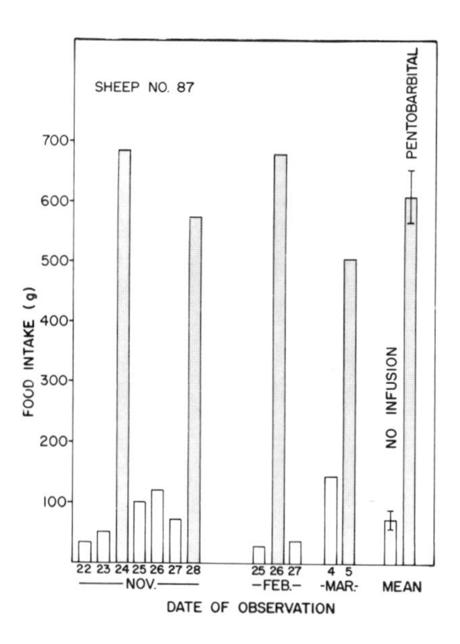

Figure 6-6. Food intake response by a sheep to infusion of the ventriculocisternal system with a depressant. (Baumgardt and Peterson, 1970)

single days, or even on groups of days. Yet it is equally apparent that these errors are corrected in the long run if energy balance (or a set rate of change in energy balance) is maintained.

Research reports dealing with the short-term control of food intake present a voluminous body of literature. A complete review is not attempted here. From the existing data it must be concluded that even within an animal species there are several feedback components which may change in relative importance depending upon physiologic and environmental conditions. It is difficult to decide whether some factors involved in the regulation of food intake should be classified as normal components of the meal feedback loop or whether they should be classified as "limiting" (or modifying) factors which inject their effect prior to the ultimate decision concerning feeding behavior.

Metering in Mouth and Pharynx

Metering in the mouth and pharynx is generally considered to be of minor significance in caloric regulation. However, there is a danger that the significance of such factors may be underestimated.

Studies of rats feeding themselves by intragastric self-injection have shown that oropharyngeal sensations are not essential for regulation of intake or for the control of such quantitative parameters of feeding as frequency and diurnal periodicity (Epstein, 1967; Epstein and Teiltelbaum, 1962; Snowdon, 1969). Animals with hypothalamic lesions are often particularly sensitive to the "palatability" of the diet. Rats with lateral hypothalamic lesions are delayed in their recovery if the food cannot be tasted or smelled. The incentive of high palatability is also necessary for vigorous overeating and for maximum levels of obesity in rats with VMA lesions (Epstein, 1967). In a recent study in which rats fed themselves a liquid diet by bar-pressing it was shown that sudden removal of oropharyngeal sensations extinguished the operant response that produced food. The response could be restored but only with prolonged training. In the absence of oropharyngeal sensations both total nutrient intake and mean meal size declined 25%. The rats maintained but did not increase their body weights. These findings were interpreted as demonstrating the potent role of oropharyngeal sensations in motivating feeding (Snowdon, 1969; 1970). It has been similarly concluded that for ruminants taste and smell has more to do with the initiation of eating (i. e., motivation) than with the amount eaten (Balch and Campling, 1962).

Gut Distension

Gastric and gut distension are usually considered of minor importance in the short-term control of food intake in most laboratory animals and in ruminants receiving a highly concentrated diet (Baile et al., 1969). Yet extensive dilution of the diet will cause restrictions in energy intake of dogs (Janowitz and Grossman, 1949; Share et al., 1952), rats (Smith and Duffy, 1957; Peterson and Baumgardt, 1971a,b), sheep (Dinius and Baumgardt, 1970), and cattle (Montgomery and Baumgardt, 1965). Distension and tension receptors have been localized in the esophagus, stomach, duodenum, and jejunum (Sharma, 1967). Distension in these areas increases the electrical activity in the vagus nerve and in the VMA. Distension is more often a limiting factor in ruminants than in nonruminants, because of the bulky nature of usual ruminant diets. We also noted in our comparative studies that the nonmature rat can adjust to a greater level of dilution (lower caloric density) than the nonmature ruminant (2.75 as against 3.50 Kcal/ml)! Based on data obtained with his experimental model, Baile (1971) considered distension as a "limiting factor" rather than a normal "meal feedback." The precise classification is open to question but I chose to include it as a normal feedback, based in part on the observation of Snowdon and Epstein (1970), who proposed a model for the control of spontaneous food intake that emphasizes peripheral controls operating against a variable set-point. It was shown that spontaneous feeding behavior of rats with verified vagotomies was normal in total nutrient intake and in the precision of regulation to dietary dilution. However, meal size was reduced by 25-56% and meal frequency was increased by 125-250%. In a subsequent study (Snowdon, 1970), it was concluded that a vagally mediated signal which initiates feeding when the preceding meal has emptied from the stomach is a most important control of spontaneous feeding. The size of a meal as well as its nutritive/osmotic components were shown to determine gastric emptying time. Vagotomy accelerated the emptying of liquid meals and prematurely activated an aversive upper gastrointestinal satiety factor.

Thus, gut distension is involved in the control of food intake; low levels of distension may enhance feeding (normal motivating signal), whereas at high levels, distension serves as a safety valve (limiting factor) to inhibit feeding. On most nondiluted, low-bulk diets, however, some other feedback must be involved in the termination of spontaneous meals.

Glucostatic Component

One of the most widely studied and accepted meal feedbacks is glucose availability or rate of utilization, as proposed by Jean Mayer. Much of the work related to the glucostatic feedback was reviewed by Mayer (1964), who demonstrated that a mechanism of food intake control based on glucose utilization could be successfully integrated with energy metabolism. Briefly listed below are the types of supportive evidence obtained (Mayer, 1964; Mayer and Arees, 1968; Anand, 1967):

(1) Satiety is related to high glucose availability (arteriovenous difference, rate of utilization) whereas hunger is related to low glucose availability.

(2) Goldthioglucose produces ventromedial lesions in the hypothalamus (of some species) with resultant hyperphagia and obesity. Possible specific glucoreceptors in VMA which, in effect, monitor glucose availability are suggested.

(3) Glucagon inhibits gastric "hunger contractions" in rats and man; the effect is not apparent in rats with ventromedial lesions. Insulin, on the other hand, increases gastric contractions.

(4) Injection of glucose results in increased response in electroencephalic recordings and single unit activity in VMA, whereas the activity in LH is reduced and "other" brain areas are not affected. Blood lipids and amino acids do not appear to alter electrical activity.

(5) Infusion of 2-deoxy glucose increases food intake in rats and dogs.

These data support the idea of a glucostatic component in the control of food intake in monogastric animals. However, several different approaches have failed to implicate a glucose feedback in ruminants. Intravenous or intraperitoneal administration of glucose (Simkins et al., 1965a,b) did not decrease food intake in ruminants, and insulin-induced hypoglycemia (Baile and Mayer, 1968c) did not increase food intake. It was recently shown that ruminants are sensitive to the toxic effects of goldthioglucose but resistant to any hypothalamic-lesioning effect (Baile et al., 1970).

Volatile Fatty Acids

In ruminants a significant part of the absorbed energy is in the form of volatile fatty acids (VFA), and relatively small quantities of glucose per se are absorbed. VFA have marked depressing effects on the food intake of ruminants (Simkins et al., 1965b) even at physiological levels (Baile, 1968). Based on extensive studies on feedbacks from spontaneous meals by goats, Baile concludes that acetate receptors are located on the lumen side of the rumen (Baile and Mayer, 1968b) and propionate receptors in the portal-liver area (Baile and Mayer, 1969). Butyrate and lactate also have been shown to decrease food intake in ruminants but information on receptor sites is less clear.

Thermostatic Component

It was suggested that heat produced during and following the consumption of a meal, as either SDA (specific dynamic action) or heat from oxidation in tissues, may be involved in a thermostatic feedback (Brobeck 1948, 1960). Such a concept has the advantage, as does the glucostatic theory, of integrating the metabolism of several metabolites. Also, there is a correlation between the SDA and satiety effects of carbohydrates, protein, and fat. The thermostatic hypothesis is difficult to test directly and several general observations do not seem to fit. Exercise produces more heat than does eating and yet has no great satiety value. Hyperthyroidism results in increased SDA of food and also in hyperphagia.

Grossman and Rechtschaffen (1967) studied brain temperature in rats and cats in relation to feeding behavior. Among their findings which do not seem to support the thermostatic idea are the following: animals kept eating after peak brain temperature was reached; animals consumed normal quantities of cold food which decreased preoptic temperature; there was no correlation between food deprivation and brain temperature. Baile and Mayer (1968a) were unable to detect any functional relationships between the hypothalamic temperature and the food intake of goats. Dinius et al. (1970), also working with goats, found that hypothalamic temperature was more closely related to activity (standing versus lying, chewing) than to food intake per se. This finding is in agreement with the observations of Grossman and Rechtschaffen (1967) cited earlier. Thus, if thermostatic feedbacks exist in relation to normal meal patterns, they are not evident by changes in hypothalamic temperature. Total heat flux might be involved but we were unable to get a hint of such in observations on skin or horn temperature (Dinius et al., 1970). It should be noted that these observations do not rule out thermoregulatory influences which might be mediated via neural pathways and reflect environmental rather than central temperature.

Osmolarity

As food and water intake are usually interrelated, it has been postulated (Lepkovsky et al., 1957) that the amount of food eaten depends on how much water can be mobilized from the tissues of the body, and that this is the basis for an osmotic regulation of food intake. The depression of food intake following a gastric load of hypertonic glucose solution has been attributed by some to its colligative properties. This possibility was tested by Yin et al. (1970). Food intake of rats was measured after gastric loads of hypertonic glucose or NaCl solution were given so as to alter the composition and volume of body fluids. Glucose markedly depressed food intake but induced only a minor increase of serum osmolality and a mild cellular dehydration, whereas NaCl affected food intake much less despite a large increase of serum osmolality and a marked cellular dehydration. From these observations and a subsequent study where repeated loadings were given, it was concluded that the glucose-induced depression of food intake cannot be attributed to an osmotic effect, and that when hyperosmolarity is induced daily it soon loses its effect on food intake.

Bergen (1969) showed that food intake by sheep is severely depressed when rumen tonicity is raised above 400 mOsm by the intraruminal administration of salts. Further observations (Bergen, 1970) indicate that high ruminal tonicity either increases water movement into the rumen or increases net digesta outflow and that high ruminal tonicity markedly depresses cellulose digestion.

Such observations lead me to suggest (for further study) that osmolarity per se is not a normal meal feedback. However, there may be a tie-in with gastrointestinal fill and motility which appears to be a normal control component (be it a meal-feedback or a limiting factor). We have found with rats and ruminants that food intake can be better predicted from kilocalories per milliliter (kilocalories per gram times density) than from kilocalories per gram alone. As was developed in earlier reports (Montgomery and Baumgardt, 1965; Baumgardt, 1970), density was selected as a tentative parameter to include factors other than digestibility which influence effective fill, such as (1) rate of digestion, (2) rate of passage, and (3) space occupied in the digestive tract per unit weight of contents. In work already cited, Snowdon (1970) found with rats consuming liquid diets that the volume and nutritive/osmotic properties of the diet determine the gastric emptying time of a meal. He further showed that gastric emptying (in the rat) is necessary to produce a vagally mediated signal to initiate a new meal. In my opinion, effective fill (in some part of the digestive tract—the site may vary among species) is the common denominator on a physiological basis which reflects the influence of secondary factors such as particle size, osmolarity, density, etc.

Other physical and chemical factors, including metabolites such as amino acids, are also involved in the overall control of food intake. (These factors are not covered in this chapter; the reader is referred to review articles such as the *Handbook of Physiology*, Section 6, Volume 1, published by the American Physiological Society.)

Long-Term (Energy Balance) Control of Food Intake

The short-term control of food intake as exemplified in the initiation and cessation of individual meals is an important part of the overall control. However, it is rather inaccurate or sloppy in its response to energy balance of the animal. Therefore, it is more than simply a convenient organization in our thinking that causes us to consider the imposition of a long-term energy balance (or memory) control system. It can be argued that long-term control is the more important for effective manipulation of energy balance of a given man or animal. Nevertheless we are woefully weak in our knowledge of clear-cut mechanisms effecting long-term regulation of energy balance. Some fresh, imaginative approaches will be necessary to get the needed answers. It appears to me that we could obtain better clues to this mechanism if we had more data on meal patterns (qualitative and quantitative) with concomitant observations on metabolite, enzymatic, and hormonal changes. LeMagnen and Tallon (1968) showed that after a 12- to 24-hour fast rats make up the calorie deficit in the first 24 hours, often by a large first meal. After a 48-hour fast, the deficit is made up over three days by occasional large meals. If these data are representative, what is it that changes within the animals's neurochemical circuitry and plumbing to induce the large, compensatory meals?

Kennedy (1953) first postulated that the size of the adipose tissue mass could signal the hypothalamic areas, and suggested that the level of plasma lipids conveyed the message. The potential involvement of fat depots as the "controlled system" is supported by several types of evidence. Cohn and Joseph (1962) induced obesity by overfeeding rats by gavage. It was found that at the end of the overfeeding period the rats did not eat a normal amount of food until their weight had fallen to the control level. Rats with VMA lesions become obese but eventually regulate body weight (and presumably body fat) at a new level. If after a prolonged fast and reduction in body weight such lesioned rats are returned to ad libitum feeding, they will return to the higher but static body weight (Gray and Liebelt, 1961; Liebelt and Perry, 1967; Teitelbaum, 1967). Liebelt et al. (1965) studied changes in various fat organs of mice caused to become obese by the administration of goldthioglucose. Changes in lipid content that occurred in two specific fat organs during the development of obesity provided evidence that each of these structures had its own intrinsic rate of lipid deposition. However, the lipid content of each fat organ continued to represent a proportional relationship to the total body lipid content as the mouse became obese. When the potential lipid-storing capacity was reduced by removing the gonadal fat organs, the inguinal fat organ demonstrated a "compensatory hypertrophy" by depositing significantly greater amounts of lipid, presumably in an attempt to compensate for the reduced lipid-storing capacity of the host. They suggest that these findings tend to shift the emphasis away from the importance of the lipid-storing capacity of the (individual) fat organs in quantitatively determining body lipid content, and direct attention to the hypothalamus. The authors suggested a working hypothesis in which the anatomically dispersed fat organs are integrated into a functional total adipose tissue mass by humoral and/or neurogenic mechanisms and in turn this adipose tissue mass exerts regulatory influences on neural pathways involved in food intake regulation, primarily in the hypothalamus, by a feedback mechanism. It was further suggested that the total adipose tissue mass is quantitatively determined in part by the intrinsic level of activity of the hypothalamic centers and in part by the intrinsic levels of activity of the fat cells.

Such evidence suggests that total fat stores may be the body component which is regulated. However, because of the limited data on total body energy balance (as discussed by Reid elsewhere in this volume), the matter cannot be considered as settled. The possibility of exchange of "information" between fat stores and lean body mass cannot be ignored. In any event, caution must be used in employing gross body weight as the indicator of body energy since many environmental, dietary, and perhaps even social factors can alter body compostion and hence energy content (Reid, 1974).

If we are unsure of the regulated component we are even less sure of the feedback mechanism by which this long-term regulation is accomplished. Circulating FFA (free fatty acids) are often mentioned as a likely candidate since they tend to increase between meals and upon fasting in monogastrics (Van Itallie and Hashim, 1960) and ruminants (Simkins et al., 1965a; Jackson and Winkler, 1970) and to decrease after a meal (Simkins et al., 1965a; Van Itallie and Hashim, 1960). The turnover of FFA is high and the level can change more than five times in a few hours. In a study of the relationship of various blood metabolites to voluntary food intake in lactating ewes, Thye et al. (1970) found that FFA concentration 7 hours after the morning meal was the best single predictor of subsequent food intake, accounting for 50% of the variation. Furthermore, in comparing an ad-libitum against a restricted meal it was found that FFA levels, at the same level before feeding, declined to about the same levels at 1½ to 3 hours after feeding. However, at 7 hours after feeding, the FFA levels of the ewes receiving the restricted meal had increased to prefeeding levels whereas the levels for those receiving a normal meal were still at the minimal level. The time trend for FFA was distinct from that for acetate, propionate, and butyrate; the concentration of these acids in the blood increased less with the limited meal than with the normal meal, but both cases led to parallel declines between 3 and 7 hours after feeding.

On the other hand, studies such as those of Hales and Kennedy (1964) show that FFA levels can increase in both hunger and in hypothalamic obesity, a situation which hardly agrees with the idea of FFA as a signal of energy balance. The data supporting FFA as a signal of long-term energy balance certainly are not conclusive, but we must be careful in discounting such ideas on the basis of conditions in metabolic disorders without considering the type and ramifications of the defect in such conditions.

Lipoproteins are just beginning to be investigated in terms of possible relationships with the regulation of energy balance and food intake (Baile, 1971). Most of the lipid in plasma is found complexed to protein as soluble lipoprotein. By means of a wide variety of physical methods, the chemist has separated plasma lipoproteins into a number of subclasses or families of compounds. In many cases the physiological implications of these classes have not been fully developed. The liver is a primary tissue in the metabolism of metabolites absorbed from the gut (meal feedbacks). Lipoproteins interact in a variety of tissues, including adipose tissue and liver. Such interactions should be investigated in relation to the regulation of energy balance.

Many hormones influence fat metabolism and hence could indirectly affect energy balance. Such a phenomenon provides a good example of the complexity of the problem in identifying *the* system(s) for control of food intake. Each body function has its own set of controls and many of them may interact to form a final thread of communication between fat stores and food intake. Only a few of the hormones which have been directly implicated will be considered here.

Kennedy (1966) suggested roles for insulin and growth hormone in an integrated scheme for long- and short-term regulation, in which a postprandial signal of satiety would be related to lipogenesis as mediated by insulin and a hunger signal would be related to lipolysis as mediated by growth hormone.

Hervey (1969) postulated that some chemical signal may work on the dilution principle to provide communication between adipose tissue and the central nervous system. If such a "tracer" were partitioned between fat and body fluids a negative energy balance would decrease the amount of fat, thus increasing the plasma concentration of the tracer which in turn might influence the hypothalamus to increase food intake. Repletion of fat stores would decrease the concentration of the tracer in the plasma and thus could reverse the process. Hervey discussed steroids as potential tracers since they tend to have the necessary solubility properties (in adipose tissue and blood), and the central control of food intake responds to steroid hormones. In this connection he discussed the probability of progesterone and adrenal glucocorticoids acting as tracers. The dilution idea is interesting but has not been sufficiently tested to allow conclusions to be drawn at this time.

The parabiotic experiments of Hervey (1959) are also supportive of an important role for hormones. VMA lesions placed in one member of a pair resulted in that member becoming very obese while the "normal" twin decreased its food intake and lost weight. Hervey suggests that the lean member of the parabiotic union was responding to a humorally carried satiety signal transported across the shared circulation and that this signal is proportional in strength to the mass of fat stored in the depot of the lesioned rat. The signal must have been noncaloric, but its nature is unspecified.

Masoro (1968) suggested that the size of the adipose tissue cells is important to the initiation of a signal. According to his postulate, when the adipose tissue cells are below a critical size no information is sent to the central nervous system from the adipose tissue. When the adipose tissue cells exceed this size they stimulate receptors in the adipose tissue which send impulses along afferent fibers to the VMA, exciting the satiety center and thus reducing food intake. It is difficult to reconcile this idea with the compensatory hypertrophy among fat organs reported by Liebelt et al. (1965).

Thus we must conclude that there is good evidence for several short-term meal feedbacks which operate with a low degree of precision, and several indications of a more precise long-term control about which we have only vague ideas and postulates.

Varying Levels of Regulation—Integration of Controls

Some attempt at integration of this scheme is in order. Attention must again be drawn to the central nervous system, especially the hypothalamus, since herein lies a large part of the circuitry for integration.

Several lines of evidence support the concept of a dual mechanism in the hypothalamus involving a ventromedial inhibitory system of neurons functioning reciprocally in the control of food intake. The individual modes of action of the VMA and LH as well as the mechanism of their communication are not fully understood. Lesions in the VMA produced by goldthioglucose result in hyperphagia and obesity in certain animals, especially mice (Baile et al., 1970; Liebelt and Perry, 1967; Mayer and Arees, 1968). This finding has been taken as evidence that the ventromedial nucleus contains certain glucoreceptors, and further work (Mayer and Arees, 1968) provided anatomical evidence that the VMA glucoreceptors act as a "satiety brake" on the LH.

Recent work has shown that the LH per se cannot be divorced from a direct involvement in at least one type of glucoreceptor in the hypothalamus. Epstein and Teitelbaum (1967) have shown that rats which are eating and controlling food intake after recovery from the initial aphagia and anorexia of LH lesions do *not* eat more during insulin-induced hypoglycemia. The deficit is specific, since such animals increase food intake in response to dilution of the diet and to low ambient temperature. Conversely, the system mobilizing food intake during hypoglycemia was intact in rats with extensive VMA destruction.

One method of visualizing the role of the hypothalamus is in terms of the set-point concept. The level of body fat stores at which regulation occurs may be determined by the hypothalamus. Teitelbaum (1961) showed that obesity caused by daily injections of zinc protamine insulin for some weeks into an intact rat was followed, when the injections were stopped, by anorexia until normal weight gain was restored. Kennedy (1966) suggested that the primary regulation carried out by the ventromedial nucleus was stabilization of the fat stores, and that hyperphagia following its partial ablation was only a means to the end of reaching a new setting of the lipostatic control.

A period of increasing hyperphagia (dynamic phase) is noted after lesioning of the VMA, but it is followed by a period of stable intake (static phase) at some higher level. Animals tend to retain the elevated body weight caused by lesioning. For example, starvation of goldthioglucose-obese mice to achieve a reduction in body weight (and lipid content), is followed by a transient increase in food intake when ad-libitum feeding is resumed (Gray and Liebelt, 1961). Also, it has been reported that varying the extent of the lesion results in graded states of hyperphagia (Kennedy, 1966; Liebelt and Perry, 1967), suggesting a set-point role for the VMA.

As indicated above, most of the implications for a set-point phenomenon have been aimed at the ventromedial hypothalamus. However, Powley and Keesey (1970) suggest a parallel role for the LH. Although the interruption of feeding resulting from LH lesions is often quite prolonged, feeding usually reappears in an orderly sequence of stages (see chart on p. 325 of Teitelbaum, 1961). LH-lesioned animals in which the ability to control food intake has been regained have been called "lateral-recovered" animals. The (temporary) loss of weight during aphagia and anorexia is usually interpreted as a secondary consequence of the failure to eat. Powley and Keesey (1970) examined the possibility that the loss of weight following LH lesions reflects an active regulatory process, and that the primary effect of the lesion is to alter the level of weight regulation (i. e., the set point). In their experiments, it was noted that even rats which had apparently recovered from the aphagic effects of LH lesions maintained their body weight a level below that of controls. It was also found that reducing an animal's weight by partial starvation prior to lesioning greatly attenuates or completely eliminates the periods of aphagia and anorexia normally seen following such LH damage. The authors concluded, " . . . the set-point for weight regulation is lowered by lesions of the lateral hypothalamus. If such is the case, the normal periods of lateral hypothalamic aphagia and anorexia are not the consequence of a breakdown in the regulation of food intake due to the neural insult; rather, they are the natural outcome of an active regulation process by which body weight is brought to this lowered level of regulation."

Thus, it appears that both the VMA and LH are involved in the long-term regulation of food intake. The precise way in which these centers communicate and the manner in which communication is channeled between body fat stores and these centers is yet to be worked out. When it is, many of us will probably pause and wonder, "Now why didn't I think of that?!"

At various times concern has been expressed about the applicability of food-intake control schemes, developed largely with animals which maintain stable body weights, to farm animals which readily fatten. It should be noted, however, that there are several laboratory animal models for genetic obesity (Renold, 1968). In most of these models, such as the genetically obese, hyperglycemic mice, the primary impairment is in lipid metabolism and any hyperphagia (which is not always present) is secondary. Although certain metabolic differences are indicated, it does *not* appear to me that fattening situations, be they "regulatory" or "metabolic" (Mayer, 1964), imply the absence of regulation. In fact, VMA-lesioned animals whose body weight and food intake have been stabilized (examples of regulatory obesity) and animals during active genetic fattening (an example of metabolic obesity) still compensate for diet dilution (Baumgardt, 1970). In studies of Liebelt et al. (1965) the administration of comparable doses of goldthioglucose to yellow obese and brown lean hybrid litter-mates resulted in comparable damage in the hypothalamuses of the obese and the lean mice. However, the lean mice showed greater food consumption and weight gain than did the obese mice. The implication is that the set-point (and lipid-storing capacity, or is it the other way around?) of the obese mice was already maximal. Could this be part of the reason why it seems difficult to get dramatic increases in hyperphagia and obesity in farm animals such as swine? As a result of VMA lesions in swine, Khalaf (1969) reported that at about 15 weeks of age (8 weeks after lesioning) operated animals were consuming 27% more food and weighed 13% more than their controls. While these differences were highly significant, the responses are not as great as usually reported for rats and mice.

Lactation presents another interesting example of increased food intake, if not hyperphagia. The explanation seems simple enough; lactation represents an increased drain of metabolites and more calories from food are necessary to balance this drain. However, close examination of several relevant observations leads me to question the completeness of this explanation. Rats nursing as few as 10 pups (Peterson and Baumgardt, 1971b) and high-producing dairy cows at peak lactation usually go into negative energy balance even on a high-energy diet fed ad libitum. This state, we suggest (Baumgardt, 1970), is due to a fill limitation imposed by the high-calorie expenditure (500 Kcal/$W_{kg}^{0.75}$ for the lactating, versus 250 Kcal/$W_{kg}^{0.75}$ for the growing rat or bovine) in relation to the relatively limited ability to extract energy from food and dispose of the residue. Of course, dilution of diets in such cases results in an even more marked lack of energy regulation. It is agreed that we are still speaking only of an increased metabolite drain. However, examination of data on the long-time changes in food intake of dairy cows repeatedly shows that peak food (dry matter) intake is not reached until well past the peak in lactation performance. Later on, when pregnancy is usually concomitant with lactation in dairy cows, the ad-libitum-fed cow tends to fatten. Did she have this physiological desire and hypothalamic setting even in early lactation, but masked by the fill problem? Or has there also been a change in the set-point? Progesterone causes increased food intake, decreased energy expenditure in certain circumstances, and gain of weight and fat. Hervey (1969) suggested that progesterone production may be a physiological means of changing the setting of regulation so that fat is stored in early pregnancy. Hervey (1969) and Kennedy (1966) have postulated hormonal feedbacks for the long-term regulation of energy balance. Is it possible that production of hormones is not a direct feedback in the usual sense, but rather that the hormones induce a change in set-point? This is somewhat implicit in Hervey's (1969) "dilution hypothesis." I earlier mentioned the work of Teitelbaum (1961), in which obesity caused by daily injections of insulin for several weeks into an intact rat was followed, when the injections were stopped, by anorexia until normal weight was restored.

Summary

Pending the availability of more research information (Figure 6-1), it is becoming well documented that individual meals are initiated and terminated in relation to feedback signals. These signals are balanced against a reference signal which is the net result of a reference input in relation to the long-term feedback signal. This long-term or energy balance feedback may be a "metabolite" which is in communication with total fat stores;

it may be something like a particular lipoprotein or FFA which, in an organ such as the liver, monitors traffic flow related to energy balance. The set-point or reference input contribution to the reference signal resides in the hypothalamus. The effective set-point can be changed in relation to the level of "certain" hormones as they are partitioned between fat and blood, the production of each hormone changing in relation to the various prime physiological functions (lactation, pregnancy, etc.) with which it is concerned. Not all hormones work on set-point. For example, the influence of growth hormone and "stilbestrol" on food intake (realizing marked differences in other functions) might be explained on the basis that they favor retention of nitrogen with the same (stilbestrol) or decreased (growth hormone) amount of fat; food intake would logically increase if the extra "calories" were going into protein rather than into fat since the energy balance feedback is related to total fat stores.

Also shown in Figure 6-1 is the scheme of an ultimate controlling system which handles the net feedback in relation to environmental, motivational, social, and limiting factors. "Limiting factors" would include things such as heat load, dehydration, metabolic deficiencies, disease and stress, and perhaps palatability (Baile, 1971). This last integration does sound like a repertoire of individual but very complex and important phenomena, each having its own feedback system. But here we must stop, with the information presently available.

This presentation is adapted and updated from "Control of feed intake in the regulation of energy balance," *Proceedings of the Third International Symposium on Digestive Physiology and Nutrition of Ruminants,* Oriel Press, Cambridge, 1970. Investigations reported herein from the author's laboratories at The University of Wisconsin and The Pennsylvania State University were supported in part by Research Grants No. AM 07652 and AM 12023, National Institute of Arthritis and Metabolic Diseases, U. S. Public Health Service. Paper no. 3854 in the journal series of the Pennsylvania Agricultural Experiment Station.

References

Anand, B. K., 1967. Central chemosensitive mechanisms related to feeding. In *Handbook of Physiology, Section 6, Alimentary Canal, Volume 1, Food and Water Intake.* Am. Physiol. Soc., Washington, D.C., p. 249.

Balch, C. C., and Campling, R. C., 1962. Regulation of voluntary food intake in ruminants. *Nutr. Abstr. Rev.* 32:669.

Baile, C. A., 1968. Regulation of feed intake in ruminants. *Fed. Proc.* 27:1361.

Baile, C. A., 1971. Control of feed intake and the fat depots. *J. Dairy Sci.* 54:564.

Baile, C. A., and Mayer, J., 1966. Hyperphagia in ruminants induced by a depressant. *Science* 151:458.

Baile, C. A., and Mayer, J., 1968a. Hypothalamic temperature and the regulation of feed intake in goats. *Am. J. Physiol.* 214:677.

Baile, C. A., and Mayer, J., 1968b. Effects of intravenous versus intraruminal injections of acetate on feed intake of goats. *J. Dairy Sci.* 51:1490.

Baile, C. A., and Mayer, J., 1968c. Effects of insulin-induced hypoglycemia and hypoacetoemia on eating behavior in goats. *J. Dairy Sci.* 51:1495.

Baile, C. A., and Mayer, J., 1969. Depression of feed intake of goats by metabolites injected during meals. *Am. J. Physiol.* 217:1830.

Baile, C. A., Mahoney, A. W., and Mayer, J., 1967. Preliminary report on hypothalamic hyperphagia in ruminants. *J. Dairy Sci.* 50:1851.

Baile, C. A., Mayer, J., Baumgardt, B. R., and Peterson, A., 1970. Comparative goldthioglucose effects on goats, sheep, dogs, rats and mice. *J. Dairy Sci.* 53:801.

Baile, C. A., Mayer, J., and McLaughlin, C., 1969. Feeding behavior of goats: ruminal distension, ingesta dilution and acetate concentration. *Am. J. Physiol.* 217:397.

Baumgardt, B. R., 1970. Control of feed intake in the regulation of energy balance. In *Physiology of Digestion and Metabolism in the Ruminant* (A. T. Phillipson, ed.). Oriel Press, Cambridge, p. 235.

Baumgardt, B. R., and Peterson, A. D., 1970. Hyperphagia in sheep induced by infusion of the ventriculo cisternal system with a depressant. *Fed. Proc.* 29:760.

Bergen, W. G., 1969. Role of rumen osmolality of feed intake by sheep. *J. Animal Sci.* 29:152.

Bergen, W. G., 1970. Osmolality and rumen function in sheep. *J. Animal Sci.* 31:236.

Brobeck, J. R., 1948. Food intake as a measure of temperature regulation. *Yale J. Biol. Med.* 20:545.

Brobeck, J. R., 1960. Food and temperature. *Recent Progr. Hormone Res.* 16:439.

Cohn, C., and Joseph, D., 1962. Influence of body weight and body fat on appetite of "normal" lean and obese rats. *Yale J. Biol. Med.* 34:598.

DeGroot, J., 1967. Organization of hypothalamic feeding mechanisms. In *Handbook of Physiology, Section 6, Alimentary Canal, Volume 1, Food and Water Intake.* Am. Physiol. Soc., Washington, D.C., p. 239.

Dinius, D. A., and Baumgardt, B. R., 1970. Regulation of food intake in ruminants. 6. Influence of caloric density of pelleted rations. *J. Dairy Sci.* 53:311.

Dinius, D. A., Kavanaugh, J. F., and Baumgardt, B. R., 1970. Regulation of food intake in ruminants. 7. Interrelations between food intake and body temperature. *J. Dairy Sci.* 53:438.

Epstein, A. N., 1967. Feeding without oropharyngeal sensations. In *The Chemical Senses and Nutrition* (M. R. Kare and O. Maller, eds.). Johns Hopkins University Press, Baltimore, p. 263.

Epstein, A. N., and Teitelbaum, P., 1962. Regulation of food intake in the absence of taste, smell, and other oropharyngeal sensations. *J. Comp. Physiol. Psych.* 55:753.

Epstein, A. N., and Teitelbaum, P., 1967. Specific loss of the hypoglycemic control of feeding in recovered lateral rats. *Am. J. Physiol.* 213:1159.

Gray, G. F., and Liebelt, R. A., 1961. Food intake studies in goldthioglucose-obese CBA mice. *Texas Rep. Biol. Med.* 19:80.

Grossman, S. P., and Rechtschaffen, A., 1967. Variations in brain temperature in relation to food intake. *Physiol. and Behavior* 2:379.

Hales, C. N., and Kennedy, G. C., 1964. Plasma glucose, nonesterified fatty acid and insulin concentrations in hypothalamic-hyperphagic rats. *Biochem. J.* 90:620.

Hervey, G. R., 1959. The effects of lesions in the hypothalamus in parabiotic rats. *J. Physiol. (Lond.)* 145:336.

Hervey, G. R., 1969. Regulation of energy balance. *Nature* 222:629.

Hetherington, A. W., and Ranson, S. W., 1939. Experimental hypothalamico-hypophyseal obesity in the rat. *Proc. Soc. Exptl. Biol. and Med.* 41:465.

Jackson, H. D., and Winkler, V. W., 1970. Effects of starvation on the fatty acid composition of adipose tissue and plasma lipids of sheep. *J. Nutr.* 100:201.

Janowitz, H. D., and Grossman, M. I., 1949. Some factors affecting the food intake of normal dogs and dogs with esophagostomy and gastric fistula. *Am. J. Physiol.* 159:143.

Kennedy, G. C., 1953. The role of depot fat in the hypothalamic control of food intake in the rat. *Proc. Roy. Soc. London,* Ser. B, 140:578.

Kennedy, G. C., 1966. Food intake, energy balance and growth. *Brit. Med. Bull.* 22:216.

Khalaf, F., 1969. Appetite control. *8th Annual Animal Science Swine Day Proceedings.* University of California Press, Davis, p. 28.

Khalaf, F., Robinson, D. W., and Jackson, H. M., 1970. Phagic response to infusions in the hypothalamus. *J. Animal Sci.* 31:205.

LeMagnen, J., and Tallon, S., 1968. The effect of previous fasting on food intake in the rat. *J. Physiol.* 58:143.

Lepkovsky, S., Lyman, R., Fleming, D., Nagumo, M., and Dimik, M. M., 1957. Gastrointestinal regulation of water and its effect on food intake and rate of digestion. *Am. J. Physiol.* 188:327.

Liebelt, R. A., Ichinoe, S., and Nicholson, N., 1965. Regulatory influences of adipose tissue on food intake and body weight. *Ann. N.Y. Acad. Sci.* 131:559.

Liebelt, R. A., and Perry, J. H., 1967. Action of goldthioglucose on the central nervous system. *In Handbook of Physiology, Section 6, Alimentary Canal, Volume 1, Food and Water Intake.* Am. Physiol. Soc., Washington, D. C., p. 271.

Masoro, E. J., 1968. *Physiological Chemistry of Lipids in Mammals.* Saunders, Philadelphia, p. 246.

Mayer, J., 1964. Appetite and the many obesities. *The Annie B. Cunning Lectures on Nutrition No. 13.* The Royal Australasian College of Physicians, Sydney.

Mayer, J., and Arees, E. A., 1968. Ventromedial glucoreceptor system. *Fed. Proc.* 27:1345.

Montgomery, M. J., and Baumgardt, B. R., 1965. Regulation of food intake in ruminants. 2. Rations varying in energy concentration and physical form. *J. Dairy Sci.* 48:1623.

Peterson, A. D., and Baumgardt, B. R., 1971a. Food and energy intake of rats fed diets varying in energy concentration and density. *J. Nutr.* 101:1057.

Peterson, A. D., and Baumgardt, B. R., 1971b. Influence of level of energy demand on the ability of rats to compensate for diet dilution. *J. Nutr.* 101:1069.

Powley, T. L., and Keesey, R. E., 1970. Relationship of body weight to the lateral hypothalamic feeding syndrome. *J. Comp. Physiol. Psych.* 70:25.

Reid, J. T., 1974. Energy metabolism in the whole animal. In *The Control of Metabolism* (J. D. Sink, ed.). The Pennsylvania State University Press, University Park, p. 113.

Renold, A. E., 1968. Spontaneous diabetes and/or obesity in laboratory rodents. In *Advances in Metabolic Disorders,* Vol. 3 (R. Levine and R. Luft, eds.). Academic Press, New York, p. 49.

Share, I., Martyniuk, E., and Grossman, M. I., 1952. Effect of prolonged intragastric feeding on oral intake in dogs. *Am. J. Physiol.* 169:229.

Sharma, K. N., 1967. Receptor mechanisms in the alimentary tract: their excitation and functions. In *Handbook of Physiology, Section 6, Alimentary Canal, Volume 1, Food and Water Intake.* Am. Physiol. Soc., Washington, D. C., p. 225.

Simkins, K. L., Suttie, J. W., and Baumgardt, B. R., 1965a. Regulation of food intake in ruminants. 3. Variation in blood and rumen metabolites in relation to food intake. *J. Dairy Sci.* 48:1629.

Simkins, K. L., Suttie, J. W., and Baumgardt, B. R., 1965b. Regulation of food intake in ruminants. 4. Effect of acetate, propionate, butyrate and glucose on voluntary food intake in dairy cattle. *J. Dairy Sci.* 48:1635.

Smith, M., and Duffy, M., 1957. Some physiological factors that influence eating behavior. *J. Comp. Physiol. Psych.* 50:601.

Snowdon, C. T., 1969. Motivation, regulation, and the control of meal parameters with oral and intragastric feeding in rats. *J. Comp. Physiol. Psych.* 69:91.

Snowdon, C. T., 1970. Gastrointestinal sensory and motor control of food intake. *J. Comp. Physiol. Psych.* 71:68.

Snowdon, C. T., and Epstein, A. N., 1970. Oral and intragastric feeding in vagotomized rats. *J. Comp. Physiol. Psych.* 71:59.

Teitelbaum, P., 1961. Disturbances in feeding and drinking behavior after hypothalamic lesions. In *Nebraska Symposium on Motivation* (M. R. Jones, ed.). University of Nebraska Press, Lincoln, p. 39.

Teitelbaum, P., 1967. Motivation and control of food intake. In *Handbook of Physiology, Section 6, Alimentary Canal, Volume 1, Food and Water Intake.* Am. Physiol. Soc., Washington, D. C., p. 319.

Tepperman, J., 1968. *Metabolic and Endocrine Physiology.* Year Book Medical Publishers, Chicago.

Thye, F. W., Warner, R. G., and Miller, P. D., 1970. Relationship of various blood metabolites to voluntary feed intake in lactating ewes. *J. Nutr.* 100:565.

Van Itallie, T. B., and Hashim, S. A., 1960. Biochemical concomitants of hunger and satiety in man. *Am. J. Clin. Nutr.* 8:587.

Welton, R. F., and Baumgardt, B. R., 1970. Relative influence of palatability on the consumption by sheep of diets diluted with 30% and 50% sawdust. *J. Dairy Sci.* 53:1771.

Yamamoto, W. S., and Brobeck, J. R., 1965. *Physiological Controls and Regulations.* W. B. Saunders Company, Philadelphia.

Yin, T. H., Hamilton, C. L., and Brobeck, J. R., 1970. Effect of body fluid disturbance on feeding in the rat. *Am. J. Physiol.* 218:1054.

7

Energy Metabolism in the Whole Animal

J. Thomas Reid

A full treatment of this subject would constitute a recounting of the almost two centuries of work carried out since Lavoisier (1777) studied respiration, Lavoisier and LaPlace (1780) observed that animal heat derives from the oxidation of organic matter in the body, and Crawford (1788) employed a calorimeter to study respiration in animals. Thus, the coverage here will be restricted mainly to certain interspecific and intraspecific peculiarities of nutritional energetics: (1) the digestibility and metabolizability of energy by various kinds of animals; (2) some between-species differences in the energy requirement of maintenance and in the net efficiency with which metabolizable energy (ME) is utilized for body gain (and other body functions); and (3) the relationship to the efficiency of energy utilization of (a) certain dietary characteristics, (b) the nature of the animal, especially its gastrointestinal architecture and physical activity, and (c) certain environmental conditions.

Partition of Dietary Energy

An overview of the ranges in the partition of dietary energy by the animals of various groups is given in Table 7.1. The magnitude of the energy losses as digestive gases, feces, urine and heat, or conversely, the proportion of dietary energy put into useful body functions, depends greatly on the nutritional quality of the diet. The values shown in Table 7-1 reflect the range in the quality of diets and in the environmental conditions to which animals sometimes are exposed, as well as the range of body functions they perform. However, the lower rates of net utilization shown are either rare, periodic in occurrence, representative of one poor-quality foodstuff, or uneconomical in practice.

Table 7-1. Ranges in the Partition of Dietary Energy (DE) by Various Animals

Energy stages		*Losses and utilization of DE (%)*		
		Simple-gutted animals (e.g., pig)	*Nonruminant herbivores (e.g., horse)*	*Ruminants (e.g., cow)*
DE → G.I. tract	Gases	0-0.5	2-7	5-12
	Feces	2-40	10-70	10-60
Absorbed energy	Urine	1-3	3-5	3-5
Metabolizable energy	Heat	5-30	10-35	10-40
Net energy: maintenance; tissue gain; milk, eggs; wool, hair, feathers; conceptus; work	Net use	25-90	15-50	10-35

For Table 7-1, animals were grouped as follows: (1) *simple-gutted animals* (SGA), including man, rat, pig, dog, and chicken; (2) *nonruminant herbivores* (NRH), including the rabbit, guinea pig, and horse; and (3) *ruminants,* including sheep, goat, cow, water buffalo, and wildebeest. Many other animals are classifiable in each of these groups, but they have not been studied in this manner. These classifications of man and the common laboratory and farm animals are based chiefly on similarities of the architecture of the gastrointestinal tract, the nature and location of the agents of digestion, and the chemical nature of the absorbed products of digestion, and therefore, on distinguishing features of digestive physiology.

For example, in SGA, digestion is effected almost entirely by tissue-elaborated enzymes in the stomach and small intestines. The small intestines represent the major nutrient-absorbing sites; glucose is absorbed as the end-product of starch digestion, and very little fermentative digestion occurs in the cecum and colon, except in the mature pig fed forage as a part of the diet, as evidenced by the production of some methane (Brierem, 1939) and volatile fatty acids (VFA) (Elsden et al., 1946). Although the digestive functions of NRH are similar in many respects to those of SGA, fermentative digestion in the cecum (and colon) of the NRH is much more extensive. Though there is some question of the extent of absorption occurring dorsally (or posteriorly) to the small intestine, some NRH (e.g., guinea pig and rabbit) benefit from coprophagy. When recycling of the feces occurs as it does normally, the digestibility of energy is increased by 10 to 25% and the retention of nitrogen is improved in rabbits (Thacker and Brandt, 1955), guinea pigs (Hintz, 1969), and rats (Tadayyon and Lutwak, 1969), over the levels observed in the same animals when coprophagy is prevented. That the horse benefits from microbial action in the cecum and subsequent absorption of amino acids is indicated by the improved nitrogen retention observed when urea is added to low-protein diets (Slade and Robinson, 1969; Hintz et al., 1969). Presumably, some VFA resulting from the fermentation of carbohydrates in the cecum also are absorbed by the NRH.

In ruminants, fermentative digestion is dominant; its major site (viz., reticulo-rumen) is ventral (or proximal) to the true stomach (abomasum) and small intestines, and absorption occurs from both the rumen and small intestines. Accordingly, the host animal is availed of certain benefits of microbial digestion and synthesis. VFA are the absorbed products of the digestion of starch and other carbohydrates. The small amount of starch escaping digestion in the reticulo-rumen is absorbed as glucose from the small intestine, as are amino acids resulting from the hydrolysis of microbial protein synthesized in the rumen.

On the other hand, microbial degradation of certain nutrients of high nutritive value is detrimental to the protein or energy economy of the ruminant. For example, when casein was administered into the rumen of sheep, its biological value was considerably lower than when it was introduced into the duodenum (Cuthbertson and Chalmers, 1950). Circumvention of hydrolysis by microbes and of ammonia loss in the rumen by delivering the casein to the duodenum of sheep resulted in a biological value for casein approaching that of casein when ingested by the rat. The feeding of liquid diets to young ruminants effects the closure of the esophageal groove to deliver the liquid from the esophagus directly to the reticulo-omasal orifice, thus circumventing the sites of fermentation. In a recent comparison of dry and liquid diets containing one of three protein sources, an average of 27% more nitrogen was retained by growing lambs provided liquid diets than by those fed the same diets in dry form (Ørskov et al., 1970).

Also, the net efficiency with which the metabolizable energy (ME), provided as glucose, was utilized for body gain was 54.5% when the glucose was infused into the rumen, and 71.5% when it was infused into the abomasum (Armstrong and Blaxter, 1961). Despite the inefficiencies invoked by microbial action in the rumen, it is only as the result of fermentative digestion that animals can derive energy from cellulose and hemicelluloses.

The animals of the SGA, NRH, and ruminant groups are different in other respects, including their natural or usual diets. The usual diet of SGA is high in concentrates and highly refined; that of ruminants is coarse, fibrous, and usually high in cellulose. The rates of digestion and of ingestal passage through the gastrointestinal tract are decidedly greater in SGA than in ruminants. The work of digestion and the heat increments are much higher in ruminants than in SGA. In SGA, the blood sugar level is considerably higher, alimentary hyperglycemia (virtually nonexistent in the mature ruminant) is much more marked, and hypoglycemia induced by fasting is greater than that in ruminants. In all of these features, NRH are intermediate to SGA and ruminants. Not all of these characteristics of the animal, its digestive function, or its diet have been related directly to the energetic efficiency of the animal. However, as suggested by Table 7-1, the energetic efficiency is highest in SGA, intermediate in NRH, and lowest in ruminants, when each is ingesting its usual diet and all are performing the same body function.

Nevertheless, not all animals are conveniently accommodated by the classification criteria employed here. Some herbivores have a fermentative-digestion device at the anterior end of the tract that is characterized usually, but not always (the kangaroo is an exception) by a mucosa of stratified squamous epithelium and a lack of glandular epithelium, by ingesta of a pH of 5 or higher, and by the presence of rumen-like bacteria and VFA. The fermentative device has different forms in various animals. In the meadow vole *(Microtus pennsylvanicus)* it is an esophageal sac (Golley, 1960); in the golden hamster *(Mesocricetus auratus)* it is a pregastric pouch in the cardiac part of the stomach (Hoover et al., 1968); and in the hippopotamus *(Hippopotamus amphibius)* it is the anterior part of the stomach which has folds and two diverticula that are also located anteriorly (Moir, 1965). In other animals such as the three-toed tree sloth *(Bradypus tridactylus),* the marsupial macropods (viz., quokka, wallaby, and kangaroo) which have a compartmentalized or sacculated stomach (Moir, 1965), and the elephant, whose cardiac portion of the stomach has transverse folds (Evans, 1910), the cardiac region of the stomach also would appear to be favorable to fermentative digestion.

It can be rationalized that the fermentative device in these various animals is arranged by shape, structure, and location to delay the passage of ingesta sufficiently long, and in some of them, it is known that the conditions are favorable for some bacterial digestion to occur. That active fermentation occurs in the stomach of the hippopotamus, the three-toed tree sloth, and the marsupial macropods is further reinforced by the presence of an esophageal groove which is the means by which the liquid (especially milk) ingesta by-passes the anterior part of the stomach to circumvent fermentative digestion (Moir, 1965). The hippopotamus does not have a cecum, and the cecum in the marsupial macropods is small and simple.

Henning and Bird (1970) observed a higher concentration of VFA in the forestomach than in the gastric pouch of kangaroos *(Macropus gigantus* and *Megaleia rufa),* and that the molar proportions of VFA resemble those observed in ruminants. Earlier, Moir et al. (1956) made similar observations in the wallaby.

Although the presence of a bacterial population and VFA in the stomach of the hippopotamus and elephant appears not to have been studied, the body fat of these animals, as well as that of the marsupial macropods, is characterized by its richness in stearic acid, a signal characteristic of ruminant body fat. This finding indicates an active hydrogenation of fatty acids in the fermentative part of the stomach, because a high proportion of the fatty acids ingested by herbivores is unsaturated.

Except for a few studies of the digestibility of energy by the hamster (Matsumoto, 1955; Hoover et al., 1968) and meadow vole (Lee and Horvath, 1968; Cowan et al., 1968; Keys and Van Soest, 1970; Petrie, 1970), the partitioning of energy by the nonruminant herbivores which have a forestomach fermentative device has not been examined. Accordingly, it is not known where in the energetics hierarchy these herbivores rank relative to SGA such as the rat at the one extreme, and the ruminants at the other.

Some Reasons for Differences in Energetic Efficiency

For the animals represented by the data in Table 7-1, the interspecific peculiarities are attributable to differences in (1) the chemical and physical nature of the usual or "natural" diet; (2) the relative degree of dependency on digestion effected by tissue-elaborated enzymes or microorganisms; (3) the chemical nature and the amounts of the absorbed products of digestion; and (4) the location of the major sites of fermentative digestion relative to the major sites of absorption. The intraspecific peculiarities in energetic efficiency are the result of: (1) the dietary characteristics, including the chemical composition and nutrient balance, the physical nature of the diet (such as particle size), and the frequency of meals (in certain animals); (2) prior nutritional history; (3) the body functions being performed currently (e.g., maintenance, growth, lactation, work, etc.); (4) the extent of voluntary or involuntary physical activity; and (5) the ambient environment, especially the temperature, humidity, altitude, rate of air movement, etc.

A series of examples will demonstrate the differences among kinds of animals, and the influence of dietary characteristics, nature of the animal, and environmental conditions upon energetic efficiency. To examine these relationships and effects, the energetics of the whole animal will be considered as those portions concerned with digestibility, metabolizability, and postabsorptive utilization of energy.

Digestibility and Metabolizability of Energy

The data generalized in Table 7-2 represent the ranges of digestible energy (DE) values, as percentages of the dietary gross energy (GE), and the metabolizable energy (ME) values, as percentages of the DE values, of good-quality diets consisting of usual or "natural" ingredients for the animals of the three groups described above. The NRH included herbivores with fermentative apparatus located dorsally to stomach and small intestines, some of which are coprophagic; the ruminants studied were mainly cattle, sheep, and goats. For NRH and ruminants, the wide range of DE values reflects the great scope in nutritional quality of their diets, ranging from fair-quality forages to those consisting of only concentrates.

Table 7-2. Digestibility and Metabolizability of the Energy of Their Usual Diets by Animals of Different Gastrointestinal Architectures

Energy ratio	*Simple-gutted animals*	*Nonruminant herbivores*	*Ruminants*
(DE/GE) × 100	75-98	40-95	50-87
(ME/DE) × 100	92-98	85-94	78-87

Digestibility of Energy of Concentrates

Although the energy of concentrates (e.g., cereal grains, oilseeds, pulses, meat and fish products, etc.) is highly digested by the animals of the three groups, it is least well digested by the ruminants. Reflected is the relatively greater efficiency of digestion by tissue-elaborated enzymes in SGA and NRH and the greater energetic expense of the digestion of readily available nutrients (such as those in concentrate foodstuffs) by microorganisms in the rumen.

Digestibility of Forage Energy

On the other hand, the digestibility of the energy of forages is controlled chiefly by the chemical nature of the plant cell walls, especially their content and distribution of lignin, hemicelluloses, and cellulose, and by the enzymatic activity of the gastrointestinal bacteria. The nutrients of the soluble contents of the cell, which otherwise could be digested by the tissue-elaborated enzymes of all mammals, are protected from digestion at least partially by the cell-wall substances, except in those animals which have an effective gut flora. Accordingly, the ruminants are best equipped to digest forages, and the digestion occurs at the proximal end of the gastrointestinal tract (an architectural feature that is detrimental to the absorption of the energy provided by concentrates). The digestibility of the plant-cell walls by NRH (even by those which are coprophagic) is somewhat less than that by ruminants, and the digestion occurs in the cecum, probably reducing the opportunity for absorption of the end products. With few exceptions (such as the mature pig, conditioned to the ingestion of forages, in which the cecum and its flora have become reasonably well developed), SGA are virtually unable to derive any energy as the result of digesting plant-cell walls. As a consequence, the "normal" diet of SGA is relatively highly refined and concentrated, and thus it has a low proportion of plant-cell walls.

Effect of Level of Input per Unit of Time

As the level of energy input provided by a diet of constant composition is increased per unit of time (e.g. per day) the digestibility of energy varies in a manner associated with gastrointestinal architecture (Reid and Tyrrell, 1964). In SGA, level of input has very little effect upon the digestibility of energy. However, increasing intakes of highly refined diets, such as one of whole milk or those composed of purified highly available ingredients, are accompanied by small increasing rates of digestibility. This response probably reflects a gradually diminishing contribution of metabolic fecal products to the total fecal matter. Nevertheless, practical diets fed to the farm animals among the SGA are seldom so highly refined, and usually will contain 2 to 10% of crude fiber. Two such diets, containing approximately 2 and 5.5% of fiber, were fed to sows at each of three levels of energy input (approximately two, four, and six times the maintenance requirement) by Parker and Clawson (1967). The digestibility of energy at the intermediate and highest levels was the same within diet (87.1 and 77.0% for the respective diets). However, at the lowest level of input, the digestibility was 89.3 and 79.3% for the corresponding diets. Thus, the digestibility of energy was depressed by only 2.5 and 3.0% as the intake of the two diets was increased from about two to six times the maintenance level.

As the level of intake of concentrate-hay diets by NRH and ruminants is increased, the DE value of the diet progressively declines. The influence of level of input upon the digestibility of energy provided to ruminants by all-forage diets is influenced by the physical nature of the forage. The digestibility is unaffected or very little affected when the forage is fed in long or chopped form (i.e., with a particle size of 1 cm or longer); however, the DE value diminishes with increasing inputs when finely ground forage is ingested by ruminants as pellets, meal, or gruel. These conclusions are abundantly documented by the literature as analyzed by Andersen et al. (1959) and Reid and Tyrrell (1964).

When the same diet is fed to cattle *(Bos taurus),* sheep, and goats, the digestibility of energy is very similar for the three species. Nevertheless, energy digestibility by some other ruminants is different from that by these species. In comparisons with sheep fed each of three diets, alpaca *(Lama pacos),* a modified ruminant with little sacs lined with glandular epithelium in the rumen, digested 10 to 50% more of the energy (Fernandez-Baca, 1966). The digestibility of the energy of nine forages, each fed singly, was 9% higher as an average for water buffalo *(Bos bubalis)* than for Holstein cattle *(Bos taurus)* (Johnson, 1966).

Differences among Animals in the Metabolizability of Energy

Table 7-2 also shows the ME value, as a percentage of the DE value, for the three groups of animals. These values are highest in SGA, intermediate in NRH, and lowest in the ruminants. The difference between these values and 100% represents the proportion of the DE (i.e., dietary GE minus fecal energy) that is lost as urine by SGA and as urine and digestive gases (mainly CH_4) by NRH and ruminants.

In SGA, the ME/DE value decreases as the amount of protein in the diet is increased above their minimum requirement, and to the degree to which the assortment of essential amino acids reaching the sites of protein synthesis is discordant with the complement of amino acids required for the protein being synthesized (Swift et al., 1957; Hamilton, 1939; Black et al., 1950). The same conditions influence the size of the urinary energy loss in NRH and ruminants, although the amino acids available for protein synthesis, relative to those provided by the diet, are modified to various degrees, depending on the extent of gastrointestinal degradation of dietary protein and gastrointestinal synthesis of protein from this source of nitrogen and nonprotein nitrogen sources. The urinary energy loss by NRH and ruminants is usually higher than that of SGA also because their diets generally contain forage. Forages provide benzoic acid, other aromatic compounds, and xylose, all of which contribute energy to the urine. Benzoic acid, upon conjugation with glycine, is excreted in the urine as hippuric acid, which has a considerably higher heat-of-combustion value (5.65 Kcal/g) than does urea (2.53 Kcal/g), the major nitrogenous excretion product of mammals, or uric acid (2.74 Kcal/g), the main excretion product of protein metabolism in birds. In the special situation of ruminants ingesting such desert shrubs as black sage *(Artemesia nova)* and big sagebrush *(Artemesia tridentata),* which contain essential oils that are excreted almost quantitatively in the urine, the urinary energy loss is quite high (Cook et al., 1952). Thus only about half the DE of these plants is metabolizable.

The loss of energy as digestive gases, especially as CH_4, is the other main reason why the ME/DE value is less than 100%. Since CH_4 loss is greatest in ruminants, intermediate in NRH, and essentially nil in most SGA, the magnitude of the ME/DE values is in the reverse order for these animals. Conditions which depress the DE value, such as increasing levels of input, also reduce the energy loss as CH_4. For NRH, the lower ME/DE value (85%) shown in Table 7-2 represents fibrous, high-forage diets; the higher value (94%) represents high-concentrate diets.

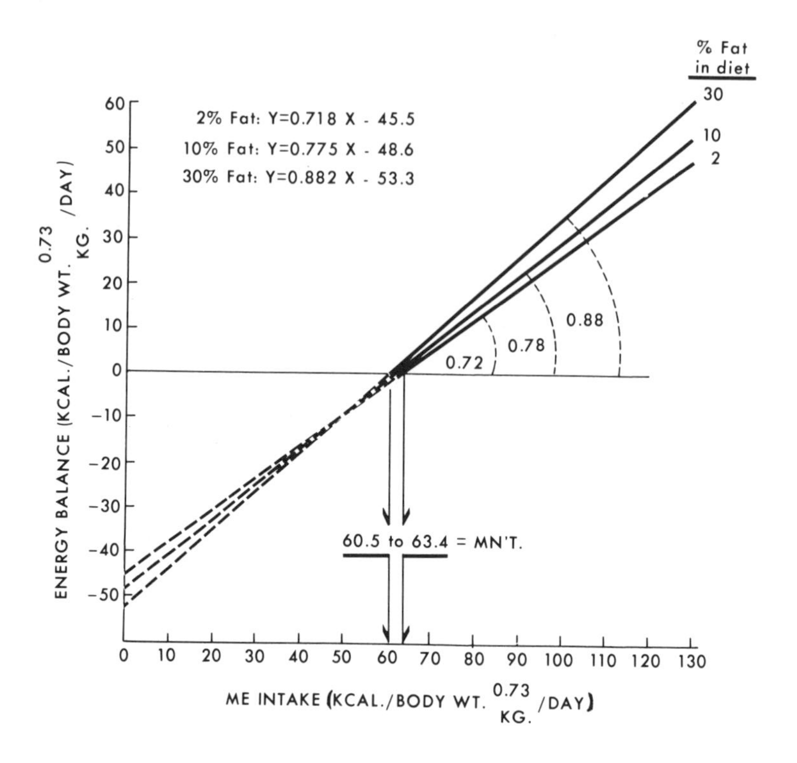

Figure 7-1. Model for examination of energetics of the whole animal.

Utilization of Metabolizable Energy by the Whole Animal

Figure 7-1 represents the model by which various kinds of influences on the utilization of ME by the whole animal will be examined here. This examination of animal energetics employs the relationship between the ME intake (X) above maintenance and the resulting energy balance (Y) —energy gain, storage, or retention. Both of these variables will be expressed per kilogram of body weight raised to the power, 0.73; this is the parameter of "metabolic body size" (MBS) that will be used. Since it is now well established, in all animals which have been studied, that the energy balance is a linear function of the ME intake above the maintenance level, the relationship can be derived with a minimum of two values. Thus, requisities of this method are the corresponding positive energy balances of two or more ME inputs within treatment. The relationship between the two variables provides: (1) the slope, the index of efficiency with which ME supports a given body function, it sometimes is called the rate of net utilization of ME for a specific body function; and (2) the ME-intake intercept at zero energy balance—i.e., the amount of ME that just prevents a gain or loss of body energy; this value represents the ME requirement of maintenance. The extrapolation of the positive energy balances to the zero ME intake derives the theoretically minimal energy balance (or, conversely, the theoretically minimal heat production as proposed by Forbes et al. (1941). In ruminants undergoing body increase as the major productive function, the energy balance extrapolated to zero ME intake usually ranges from approximately -45 to -60 Kcal/MBS/day. Since orderly study has not been made of this characteristic in other animals, the degree to which it is general among animals is not known.

The specific example shown in Figure 7-1 was adapted from the data of Forbes et al. (1946) obtained in an experiment in which 12 6-month-old male rats (weighing 311 to 350 g during calorimetric periods) were fed isonitrogenous and isocaloric diets containing 2, 10, or 30% of fat at each of two levels of GE intake (44.2 and 64.4 Kcal/day). As the level of fat in the diet was increased from 2 to 30%, the net utilization of ME for body energy gain increased from 71.8 to 88.2%. Since the diets were isonitrogenous as well as isocaloric, naturally the substitution of fat in the diet required small increments of protein and marked decrements of carbohydrate as the fat level was increased. However, the maintenance requirements ranged only from 60.5 to 63.4 Kcal of ME/MBS/day for the rats ingesting the three diets. These values are considerably lower than those usually observed in young, active animals. In fact, these values are lower than the interspecific fasting metabolic rate (70.4 Kcal/MBS/day) of mature animals as proposed by Brody and Procter (1932). In Brody and Procter's study, however, the rats were mature, and they were rendered immobile to the point essentially of slumber by a bright light focused on their eyes while they occupied the confining chamber of the Haldane respiration apparatus.

Energetic Efficiency of Body Gain in Various Kinds of Animals

Table 7-3 shows examples of the influence of several dietary characteristics, physical activity, and ambient environment on the ME requirement of maintenance and the net utilization of ME for body gain in certain animals. The examples are designated by experiment numbers as follows for the three groups of animals: SGA, 1-9; NRH, 10-12; and ruminants, 13-21.

Table 7-3. Energy Requirement of Maintenance and Net Utilization for Body Gain of Metabolizable Energy Ingested above Maintenance Level for Various Animals as Reported in the Literature

Experiment no.	Animal	Main treatment(s)	Other distinguishing conditions	Net utilization (% of ME)	Maintenance requirement* (Kcal/MBS/day)	Source
1	Rat (6 mo. old; 310-350 g body wt.)	Fat (%): 2	Isonitrogenous, isocaloric diets; each fed at 2 levels; voluntary activity prevented; temp., 29° C	71.7	63.4	Forbes et al., 1946
		10		77.5	62.8	
		30		88.1	60.5	
2	Rat (weanling; 48 g body wt.)	Protein (%): 10	Equicaloric amounts provided at each of 2 levels during 70-day body balance experiment; temp., 29° C	23.7	86.3	Forbes et al., 1939
		25		72.8	127.4	
		35		73.9	125.1	
		45		75.6	130.6	
3	Rat	Amb. temp. (° C): 25	5 levels of energy input provided by same diet at each temperature	76.2	(138)†	Nehring, 1958
		27.5		76.6	(115)†	
		30		75.9	(100)†	
4	Pig (initial body wt. ca. 100-180 kg)	Corn and barley	Three experiments; 2 basal diets, usual ingredients, provided at 4-6 levels of energy input	73.6	88.6	Nehring et al., 1960b
		Barley and oats		81.1	116.1	
		Barley and oats		76.4	101.9	
5	Man 1 (62 kg)	Two levels of energy input	Subjects were underweight; 21-24 years old; in bed from 8 P.M. to 1:30 P.M.; walked 5 mi./day at rate of 4 mi./hr.	96.0	114.2	Passmore et al., 1955
	Man 2 (59 kg)			91.0	107.0	
	Man 3 (57 kg)			90.0	110.0	
6	Woman (21 years old, 129 kg)	No exercise	Obese subjects; each of 3 levels of energy input imposed during 4- to 8-wk. periods	89.2	73.9	Buskirk et al., 1963
		Exercise		84.8	92.0	
	Man (19 years old, 145 kg)	No exercise		78.7	72.8	
		Exercise		76.1	99.2	

7	Chicken (10-25 days old)	Protein (%): 6.3	4 levels of energy input imposed on each level of protein; 15-day body balance experiment	64.2	170.9	Thomas, 1966
		12.6		68.7	177.3	
		18.9		79.5	172.0	
		25.2		80.3	169.8	
8	Chicken (mature, 3.5-4.8 kg)	4% fat, 28° C	4% fat diet contained 80% corn meal; 32% fat diet contained 29% corn oil	73.0	84.6	Shannon and Brown, 1969
		4% fat, 22° C		69.2	99.8	
		32% fat, 22° C		84.3	95.2	
9	Calf (*Bos taurus*) (16-50 days old)	Whole milk diet	2 levels of input	84.6	144.1	Blaxter, 1952
10	Rabbit (9-12 mo. old)	6 levels of input above maintenance	Diet of 60% wheat + 40% hay; temperature, 17.8-22.8° C	70.5	106.5	Hellberg, 1949
11	Guinea pig (130-800 g body wt.)	3 rates of growth	3 levels of energy input; 28-30° C	69.6	86.8	Knudsen, 1968
12	Guinea pig (6-8 mo. old)	10.9% protein, 20° C	Diet of 45% roughage, 55% concentrates; each diet fed at each of 4 levels of input	66.8	84.2	Vercoe, 1965
		24.9% protein, 20° C		66.2	81.8	
		19.1% protein, 24° C		63.6	68.0	
		30.9% protein, 24° C		70.9	69.2	
13	Sheep (15 mo. old)	Long hay, 5.7% lignin	6 levels of input	41.9	95.8	Blaxter and Graham, 1955
		Pell., gr. hay, 7.7% lignin	8 levels of input	51.7	88.9	
14	Sheep (mature)	Chopped hay	Same forage in all diets; 2 levels of input of each diet	36.3	92.6‡	Blaxter and Graham, 1956
		Pell., med. gr. hay		48.7	87.2‡	
		Pell., finely gr. hay		44.2	83.5‡	
15	Sheep (8 and 20 mo. old at beginning)	Chopped hay	Each diet fed at 3 levels; same hay in all diets; slaughter-analysis experiment	30.8	91.5	Paladines et al., 1964
		Pell., gr. hay		43.4	86.7	
		Pell., 50% hay + 50% corn		56.9	82.4	

(continued on next page)

Table 7-3 (continued)

16	Sheep (3½ year-old ewes)	40% hay + 60% concentrates	Each of 5 levels of input imposed during 10-wk. periods	62.5	87.9	Marston, 1948
17	Sheep (6-7 mo. old at beginning)	Hay + glycerol	Each diet fed at 2 levels; 175-day feeding period; slaughter-analysis experiment	59.5	86.4	Bull et al., 1970
		Hay + triacetin		59.9	88.5	
		Hay + corn + glycerol		61.8	85.1	
		Hay + corn + triacetin		63.7	87.8	
18	Cattle (steers, 2 years old, 614 kg)	24% hay + 76% concentrates	Diet fed at 5 levels of input	61.3	107.0	Mitchell et al., 1932
19	Cattle (steers, 2-3 years old, 490 kg)	50% hay + 50% corn	Diet fed at 4 levels of input	51.0	102.0	Forbes et al., 1928, 1930
20	Steers, mature	66% hay + 34% oats	Each steer fed at each of 5 levels; each wether fed at each of 6 levels	54.6§	106.4	Blaxter and Wainman, 1961
	Wethers, mature			53.6§	74.4	
21	Wildebeest (*Connochaetes* (*taurinus*). mature	ca. 50% hay + 50% concentrates	Diet of this composition fed at 2 levels of input	59.2	Body wt. not reported.	Rogerson, 1968

*Maintenance requirement expressed as kilocalories of metabolizable energy (ME).
†Since body weights were not reported, the ME requirement is expressed as a percentage of that at 30° C.
‡Body weights were not reported; to estimate the requirement, it was assumed that the mean body weight is 37 kg.
§Derived from the authors' data for positive balances; MBS values were computed from the fasting body weights.

Simple-gutted animals (SGA). The data of Forbes et al. (1946), obtained in the study of the effects of fat concentration in the diet as cited above and employed to construct the diagram in Figure 7-1, are restated as Experiment 1 to establish the format employed in Table 7-3.

Experiment 2 was conducted by Forbes et al. (1939) with young rats weighing 48 g at the beginning. The rats were quadruple-fed each of two levels of intake of diets providing 10, 25, 35, and 45% of protein, as beef muscle, during 70-day periods. The beef muscle was substituted in the diet equicalorically for dextrin and Crisco. Energy storage was determined by analyzing the bodies of littermates at the beginning and those of the experimental animals at the end of the feeding period, and the ME intake was determined by measuring the energy loss in the excreta over the 70-day period. The net utilization (72.8 to 75.6%) of ME for body gain was similar for the rats fed the diets containing 25 to 45% of protein; for those fed the 10% diet, the efficiency of energy gain was markedly lower (23.7%). The maintenance requirement of rats fed the 10% protein diet was much lower than that of the rats ingesting the 25 to 45% protein diets, reflecting the morbidity of the low-protein-diet animals. Although the higher maintenance requirements of rats fed the 25, 35, and 45% protein diets might reflect the increased energy cost of urea synthesis, the values are not graduated in a manner that would so indicate. Nevertheless, the maintenance cost of the active, young rats in this experiment is about twice that of the immobile, mature rats used in Experiment 1.

That ambient temperature has a drastic influence on the ME cost of maintenance in the rat is demonstrated by Experiment 3, conducted by Nehring (1958). The same diet was provided at each of five levels of input at each of three temperatures (25, 27.5, and 30°C). Although the net efficiency of body gain was not different, the maintenance requirement was 15% greater at 27.5°C and 38% greater at 25°C than it was at 30°C. (Since the author did not report body weights, the values in Table 7-3 are the maintenance requirements expressed as a percentage of that at 30°C, the thermally indifferent temperature of the rat.)

Experiment 4 represents an extensive series of energy-balance experiments involving two distinct basal diets, each provided at four to six levels of input as increments of the basal diet or of barley meal to each of 4, 6, or 8 pigs as they grew from a body weight of about 100 to 180 kg (Nehring et al., 1960b). The net efficiency of body gain observed in these studies is of the same order of magnitude as that reported for other SGA. The ME requirement of maintenance was considerably higher for the pigs ingesting the diets high in barley and oat meal than for those whose diet contained 42% of corn meal with 42% of barley meal plus the same protein supplements.

Experiment 5 was conducted by Passmore et al. (1955) with three thin (underweight) but healthy men, 21 to 24 years of age. Each of two levels of input were imposed on each subject. The daily routine consisted of rising at 1:30 P.M., walking 5 miles each afternoon at the rate of 4 miles per hour on a negligible grade, eating 4 meals, and returning to bed at 8 P.M. Oxygen consumption was measured during 10-minute periods (except during walking, when the periods were 5 minutes) at 4-hour intervals throughout the day and night. The mean ME requirement of maintenance was 110 Kcal/MBS/day and the net utilization of ME for body energy gain was 92.1%.

The data designated as Experiment 6 were derived from the extensive studies made by Buskirk et al. (1963) of obese human subjects. During periods of 4 to 8 weeks, each of three levels of energy intake was provided with and without exercise on a treadmill (7.5 miles per day at the rate of 3 miles per hour on a 5% grade). However, because of chafing of the contacting body surfaces, the subjects were allowed to rest on an upright bar and did not perform the full amount of work planned. Some of the diets were constituted from a Metrecal base and others were formulated from milk solids, dextrimaltose, and corn oil and ingested as liquids. The net utilization of ME within subject was not influenced by exercise; however, the ME requirement of maintenance was decidedly increased by exercise. A comparison of the maintenance requirements of the thin men (Experiment 5) with those of the obese subjects (Experiment 6) indicates that those of the latter are much lower, but conclusions cannot be drawn until such studies are conducted under the same conditions of diet, ambiency, and experimentation.

In Experiment 7, Thomas (1966) fed 10-day-old chickens each of four levels of protein superimposed on each of four levels of energy input. Energy storage was determined by chemical analysis of the whole body of representative chicks at the beginning and of the experimental birds after a 15-day feeding period. As shown in Table 7-3, the maintenance requirements were not different for the protein-level groups, and therefore did not graduate in a manner that would suggest an increasing energy cost of uric acid synthesis with increasing dietary protein level. However, the net efficiency of body-energy gain increased as the protein level in the diet increased to a level of 18.9%. Since the net utilization of ME increased approximately linearly as the protein level increased from 6.4 to 18.9%, and no perceptible additional increase was effected by the 25.2% protein diet, the protein requirement under the conditions of this experiment was of the order of 18.9% of the diet. A high ME requirement of maintenance was found for these very young (10- to-25-day-old) birds.

In Experiment 8, Shannon and Brown (1969) examined the influence of dietary fat level (approximately 4 and 32% of dry matter) at an ambient temperature of 22°C, and that of ambient temperature (22 and 28°C) when the 4% fat diet was fed to mature Light Sussex cockerels weighing 3.5 to 4.8 kg. Energy exchange was derived from the continuous measurement of oxygen consumption and carbon dioxide production during 22 hours of each 24-hour metabolism period. The remaining 2 hours were employed for excreta collection and refeeding. Each bird was maintained on a given intake level for 7 days, the last 3 of which were spent in the respiration calorimeter. At least 4 weeks were allowed for acclimatization at either temperature.

At 22°C, the maintenance requirement was about the same for the birds fed the two diets, the net utilization of ME for body gain was about 22% greater for the birds ingesting the 32% fat diet than for those receiving the 4% fat diet. This result is similar to that recorded for the rats employed in Experiment 1 (Forbes et al., 1946) and that observed by Chudy and Schiemann (1967), also with rats. Increasing proportions of fat in the diet have increased the rate of energy storage in sheep as well (Swift et al., 1948), though the effect was not as dramatic as those of Experiments 1 and 8. In an experiment conducted with two pairs of young men ingesting diets containing 4.9 and 34.5% of fat (provided by butter and heavy cream), Swift et al. (1959) found no effect of fat concentration on the efficiency of body-energy storage. Except in the latter experiment, it appears that dietary fatty acids are used preferentially to form body fat. The higher energetic efficiency of high-fat diets undoubtedly results from the direct incorporation of dietary fatty acids into triglyceride. This direct incorporation circumvents the energy cost of denovo synthesis of fatty acids from other dietary sources, a process requiring reduced nicotinamide adenine dinucleotide phosphate.

The chickens employed in Experiment 8 (Table 7-3) that were fed the 4% fat diet required more ME for maintenance at the ambient temperature of 22°C than at 28°C. However, the net efficiency of body-energy gain was not different for the two temperatures.

In Experiment 9, conducted by Blaxter (1952), two levels (4 and 8 liters per day) of cow's whole milk were imposed on nonruminating calves beginning at 3 days of age; ruminants of this age are devoid of gastrointestinal flora and therefore are essentially SGA. At intervals between 16 and 50 days of age, heat production was determined by means of respiration calorimetry. The energy requirement of maintenance for the young calf is high, as is that of the young rat (Experiment 2) and chicken (Experiment 7). In contrast to that of the mature bovine (Experiments 18, 19, and 20), the net efficiency of body gain in the young calf ingesting whole milk is very high. Liquid diets ingested by young ruminants are transported via the esophageal groove to the reticulo-omasal orifice, thereby avoiding fermentation in the reticulo-rumen. In this study, the metabolizability of the GE of the milk was 95.5%. Thus, the net utilization for body gain of the GE ingested above maintenance was 81% [i.e., $(0.955 \times 0.846)100 = 81$].

Nonruminant herbivores (NRH). Experiment 10 was conducted with mature rabbits weighing approximately 3 kg (Hellberg, 1949). A diet consisting of 60% of coarsely ground wheat and 40% of ground hay was fed at six levels above that of maintenance. Energy storage was determined by the Haldane method at ambient temperatures ranging from 17.8 to 22.8°C, with most balances being determined at a temperature below 20°C. Whether the maintenance requirement would have been lower at a higher ambient temperature than that determined (i.e., than 106.5 Kcal of ME/MBS/day) is not known. There is much disagreement about the temperature range of thermal comfort (i.e., the range of minimal heat production) for the rabbit. For example, Tomme and Misiutkina (1936) reported the range to be 15 to 20°C, but Lee (1942) observed that the lower critical temperature might be as high as 28°C. It is probable that differences in the nature and amount of the fur, and the extent of adaptation of the rabbit to the various environments, may have affected the disparate critical temperatures observed.

Experiments 11 and 12 concern guinea pigs. In Experiment 11, young guinea pigs (10 to 40 days old), weighing 110 to 300 g at the beginning, were fed continuously at one of three levels of energy input during periods of 1 to 4 months under an ambient temperature of 28 to 30°C (Knudson, 1968). Heat production was determined by the Haldane method when the animals weighed from 130 to 750 g. The guinea pigs employed in Experiment 12 (Vercoe, 1965) were 6 to 8 months old and weighed 400 to 765 g. Diets containing 10.9 and 24.0% of protein were fed at an environmental temperature of 20°C, and diets containing 19.1 and 30.9% of protein were fed at a temperature of 24°C. All diets contained 57.9% of a mixture of crushed wheat and oats, oat straw, alfalfa hay, and mineral and vitamin supplements; the remaining 42.1% was constituted by four different proportions of wheat gluten and cornflour starch to provide the four concentrations of protein. Each diet was imposed at four levels of input and the energy balance was determined by the Haldane method. As shown in Table 7-3, the net utilization of ME for body gain was not different among the protein levels. Although the ME requirement of maintenance was not influenced by the protein concentration of the diet, the data for the maintenance requirements segregated in a manner associated with the ambient temperature. The requirements determined at 20°C were about 21% higher than those determined at 24°C. (The author indicated that the temperature inside the Haldane chamber was about 3°C higher than that in the surrounding room; therefore, the ambient temperatures employed in this experiment were probably of the order of 23 and 27°C).

Ruminants. Experiments 13-21 were selected to demonstrate certain interspecific peculiarities in, and the usual range of nutritional influence on, the energetics of ruminants. These data (Table 7-3) indicate that the net efficiency with which ruminants utilize for body gain the ME ingested above the maintenance level usually ranges from about 30 to 65%. However, the efficiency can be as low as 10% when the diet consists of a highly lignified forage such as wheat straw. When body gain occurs simultaneously with lactation in the goat (Armstrong and Blaxter, 1965) and the bovine (Moe and Flatt, 1969; Moe et al., 1970), the net efficiency of body-energy gain may be as high as 70%.

For fattening ruminants (chiefly male castrates), the ME requirement of maintenance ranges from approximately 80 to 110 Kcal/MBS/day. Sheep appear to have a lower maintenance requirement than do cattle. For other ruminants the data are insufficient to determine how their energetics compare with these two more commonly studied species. In studies not cited in Table 7-3, the maintenance requirement of nonlactating, nonpregnant cows has ranged from about 110 to 135 Kcal/MBS/day (van Es et al., 1961, 1965; van Es and Nijkamp, 1967a; Moe et al., 1970). The maintenance requirement of lactating cows (138 Kcal) was about 22% higher than that (113 Kcal) of nonlactating cows (Moe et al., 1970); however, van Es and Nijkamp (1967a,b) detected no difference between lactating and nonlactating cows.

Experiments 14 and 15 (Table 7-3) show that when pelleted finely ground hay is ingested by sheep, the net utilization of ME for body gain is greater and the ME requirement of maintenance is somewhat less than when hay from the same source is fed in chopped form. The addition of corn meal to the hay employed in Experiment 15 resulted in a further improvement in efficiency, and some reduction in the maintenance cost. In his classical experiment (Experiment 16), carried out with 3.5-year-old Merino ewes fed a diet consisting of 50% crushed wheat, 10% cane molasses, and 40% ground alfalfa hay, Marston (1948) observed a net efficiency of 62.5% and a maintenance requirement of 87.9 Kcal of ME/MBS/day. Bull et al. (1970) obtained similar averages (Experiment 17, Table 7-3) for male and female Southdowns (both with intact gonads) fed a diet of ultra-high-quality ground alfalfa hay and corn meal, supplemented with 1.8% of glycerol or 8.4% of acetic acid and 1.8% of glycerol. The net efficiency with which the ME provided by the same diets lacking the corn meal, was used for body energy gain was only a little lower.

Corresponding data for cattle fed mixed concentrate-hay diets are labeled Experiments 18 and 19 in Table 7-3. The most conspicuous characteristic of these data, as compared with those for sheep (Experiments 13-17), is the higher ME requirement of maintenance. In Experiment 20, Blaxter and Wainman (1961) compared adult wethers and steers directly. They fed a diet of two parts poor-quality hay and one part oats at each of five levels of energy input to each of 3 steers weighing 380, 460, or 490 kg during fasting, and at each of six levels to each of 3 wethers weighing 39, 55, or 63 kg during fasting. As shown in Table 7-3, the net utilization of the ME ingested above maintenance level was not different for the two kinds of animals, but the ME requirement of maintenance was 74.4 and 106.4 Kcal/MBS/day for the wethers and steers, respectively; the latter values were computed for the present study with the fasted body weights used to determine the MBS. The net utilization of ME for maintenance was 80.4% in both species.

In another experiment (not cited in Table 7-3), Blaxter and Wainman (1964) compared 3 wethers with 3 steers each fed each of five diets at each of two levels of intake. The diets consisted of a mixed hay and cooked, flaked corn in various proportions (see Table 7-4). Although these data again demonstrated that the net efficiency of fattening is the same in sheep and cattle ingesting the same diet and performing the same body functions, they also demonstrate the age-old phenomenon that the ME provided by concentrates is considerably more productive than that supplied by forage.

Table 7-4. Rates of ME Utilization for Body Gain for Sheep (Wethers) and Cattle (Steers) on Varying Diets (%)

Diet mixture (%)			
Hay	*Corn*	*Sheep*	*Cattle*
100	0	27.5	31.4
80	20	35.4	32.8
60	40	44.0	42.1
40	60	47.1	47.6
20	80	53.3	54.2
5	95	—	63.5
0	100	58.8	—

The net utilization of ME for body gain by a wildebeest, an African ungulate, is recorded as Experiment 21. Because of their greater leanness of carcass, some of the wild ruminants have been considered to be more efficient energetically than domesticated ruminants. However, at least when ingesting the domesticated ruminant's diet, the wildebeest appears to utilize ME for body gain at about the same rate as do domesticated ruminants. Unfortunately, Rogerson (1968) did not report body weights and, therefore, it is not possible to determine the maintenance cost per unit of MBS.

Relation of age, chemical nature of synthesized tissues, and sex to energetic efficiency. Whether the efficiency with which ME is utilized for body gain differs between the young animal and the mature animal is unsettled. Since the body substance gained by the mature subject is much higher in fat and lower in protein than that of its young counterpart, the question of whether the efficiency of fat synthesis is different from that of protein synthesis is associated with the question of whether age is associated with energetic efficiency. Based on theoretical considerations, it appears that the energetic efficiency of protein synthesis (90 to 93%) or of glycogen synthesis (97%) is high compared with that of fat synthesis (70% or less) (Blaxter, 1962; Schiemann, 1963). As the result of slaughter experiments with lambs and pigs, Kielanowski (1965) obtained data which indicate an efficiency of ME utilization of 80 and 75% for protein synthesis, and of 63 and 81% for fat synthesis, in the respective animals. On the other hand, Thorbek's experiment (1967) with pigs indicated that the efficiency of ME utilization above maintenance level was markedly lower when the proportion of the total energy gain as protein was high (i.e., the proportion as fat was low) than vice versa. In these experiments, in which it was the intent to estimate the relative efficiency of protein and fat formation from the ME available above maintenance level in the whole animal, the ME cost of maintenance was determined by means of equations such as that of Breirem (1939) for the pig, viz.:

$$\text{ME (Kcal/day)} = 196.3 \times \text{body weight}_{\text{kg}}{}^{0.56}$$

However, a more valid estimate of the efficiency of protein and fat formation could have been made if two levels of energy input had been imposed in order that the maintenance requirement could have been determined directly at each stage of growth.

Age, within the specified limits studied, and its associated difference in the chemical composition of body gain appear not to be related to the energetic efficiency of sheep. In the experiment of Schürch (1961), in which 3- to 4-year-old wethers were compared with 6- to 15-month-old wethers, and in that of Paladines et al. (1964), in which wethers 8 months old at the beginning of a 7-month feeding period terminated by slaughter and whole-body analysis were compared with 20-month-old wethers also exposed to the 7-month feeding period, neither the maintenance requirement nor the efficiency of utilization of ME ingested above maintenance level was different for the two age groups. In the latter experiment (Paladines et al., 1964), and probably in the experiment of Schürch (1961), the body gain of the older sheep consisted of a markedly higher proportion of fat and a lower proportion of protein than did that of the younger animals.

In other situations, a high-fat gain has been associated with a high energetic efficiency. For example, the body gain of rats fed ad libitum contained 58% of water, 27% of protein, and 13% of fat; whereas that of their pair-fed mates receiving the same amount of energy per day in two meals via stomach tube contained 29% of water, 22% of protein, and 44% of fat (Han and Reid, unpublished data). The body-energy gain expressed as a percentage of the total ME intake (i.e., including the cost of maintenance) was 12.2 for the ad libitum-fed rats, and 23.5 for those given two meals per day.

Bull et al. (1970) examined the body composition of male and female sheep (both with intact gonads) at the beginning and end of a 175-day feeding period during which one of four different diets was provided to a given animal. Although the females contained from 30 to 35% more fat and 3 to 17% less protein at any body weight from 20 to 50 kg (on the ingesta-free basis) than the males, and over that body-weight range, females gained 30% more fat but 31% less protein, the females retained 65.5% of the ME ingested above the maintenance level, as an average for the four diet groups, whereas the males stored 57.6%. Despite this difference in efficiency, the ME requirement of maintenance was the same for the two sexes (i.e., 88 and 87 Kcal of ME/MBS/day for the males and females, respectively).

The disparity between the results obtained in the whole animal, as cited here, and those based on theoretical considerations demonstrates that current biochemical information either is inadequate in itself, or has been inadequately fitted to the whole-animal system. In either case, it is imperative that future research of whole-animal calorimetry simultaneously include studies of cellular biochemistry—and, where feasible, in the same subject. In this context, protein synthesis would seem to have cardinal significance.

Comparative energetics of animals. As exemplified by the data in Table 7-3, the net efficiency with which the ME ingested above the maintenance level is utilized for body gain by animals ingesting their usual practical diets ranges as follows for the three groups of animals considered here: SGA, 75 to 90%; NRH, 60 to 70%; and ruminants, 30 to 65%. Although the magnitude of this value is not affected by physical exercise and ambient environment, dietary deficiencies of protein and of certain mineral elements and vitamins, and an imbalanced assortment of amino acids impair the utilization of ME ingested above the maintenance level. On the other hand, the ME requirement for maintenance is increased by increasing physical activity (both voluntary and involuntary), an ambient temperature that is higher or lower than the zone of thermal neutrality, and a high altitude (e.g., 14,110 ft.) (Chinn and Hannon, 1969).

The maintenance requirement per unit of body weight raised to the power, 0.73, is high in the young animal and recedes to a lower but relatively stable value in the mature subject, apparently in all homeotherms. For example, Graham (1966) observed a fasting metabolic rate in adult sheep (1.5 to 6 years old) of 57 Kcal/MBS/day; in lambs 5 to 13 weeks old, the fasting metabolism was 100 to 104 Kcal/MBS/day. The interspecific fasting metabolic rate of mature animals is generally considered to be approximately 70 Kcal/MBS/day, as proposed by Brody and Procter (1932). However, Blaxter (1962) has called attention to the fact that the fasting metabolic rate of sheep is lower (as indicated also by Graham's data (1966) above) and that of mature cattle (78 to 96 Kcal/MBS/day) is higher, as shown also by the data of Blaxter and Wainman (1961, 1964, 1966), than the interspecific mean of Brody and Procter (1932). That the maintenance requirements of these two species are similarly different is borne out by the data in Table 7-3.

The fasting metabolism of bulls of the Brahman, Africander, and Hereford × Shorthorn breeds, all 13, 20, or 22 months of age during a given year of experimentation, was studied by Vercoe (1970). The average fasting rates of the respective breed populations were 86.4, 102.5, and 97.4 Kcal per kilogram of body weight to the 0.75 power. Although the rate of the heat-resistant Brahmans was the lowest of the three breeds and that of the heat-resistant Africanders was the highest, the rates for the bulls of all breeds were high relative to other data from similar studies reported in the literature. Some of the reasons for the differences might be a prior ad-libitum feeding level, the relatively short-term (7-hr) measurements of heat production during which the animals were generally in the standing position, and the sex of the subjects.

The basal metabolic rate of some Australian mature marsupials is also considerably lower than the interspecific mean rate of 70 Kcal/day/MBS. In eight species ranging in size from the white-tailed marsupial mouse (9.0 g body weight) to the red kangaroo (54 kg body weight), Dawson and

Hulbert (1970) observed an average fasting metabolic rate of 51.3 Kcal/MBS/day. Thus it is indicated that the energy requirement of maintenance might be 25 to 30% lower in these animals than it is in the eutherian mammals.

It seems probable that a given constant parameter such as body weight to the power 0.73 may not be adequate for all comparisons made between, or even within, species at all ages. For example, from the data of Cairnie and Pullar (1957), the body-weight exponent of best fit was computed to be 1.32 for heat production by very young pigs fed ad libitum at 30°C as they grew from a body weight of 4 to 12 kg. That study, carried out at four temperature intervals from 15 to 30°C, indicates that the critical temperature is successively higher than 30°C for pigs smaller than 10 kg, and may be as high as 40 to 45°C in the 4-kg pig. As body size increased, it is probable that the intake per unit of body weight and the extent of body insulation as hair and body fat increased. These conditions might account in part for the fact that the rate of change in heat production (1.32) per unit change in body weight is considerably greater than the rate of change in basal heat production (0.56), as observed by Breirem (1939) in older pigs. However, undoubtedly a part of the difference is the result of the higher metabolic rate of the young animal.

Although the fasting metabolic rate of some Australian mature marsupials was considerably lower than that of eutherian mammals, the body-weight exponent giving best fit to the metabolic rate of the marsupials was 0.737 (Dawson and Hulbert, 1970). This value is about the same as that (0.734) derived by Brody and Procter (1932) from the data which yielded their interspecific values.

Most workers seem content to employ body weight raised to a power ranging from 0.7 to 0.75 as the physiologically effective body size even in application to young animals, despite the fact that these parameters were derived in studies of mature animals. Nevertheless, study is needed of other reference bases, e.g., lean-body mass or body protein, potassium, or DNA, which might have greater physiological significance or which might serve as a more effective predictor of the fasting metabolic rate. Graham (1966) examined the relationship between lean-body weight and fasting heat production in sheep and compared it with that in man as observed by Behnke (1953) and Miller and Blyth (1953). The fasting metabolic rates per kilogram of lean-body weight per day were about 27 Kcal for the adults of both species and 54 Kcal for lambs before or at weaning and for children. Although it is clear that even on the lean-body basis the metabolic rates of young and mature animals are different, it is not known to what degree the metabolic rate is similar within age group among various other animal species.

The cause of the lower energetic efficiency of ruminants than of SGA has been sought but has remained a mystery for almost a century. Von Mering and Zuntz (1877) and Kellner (1905) attributed the ruminant's high heat increment to the "work of digestion." Although the relatively low energetic efficiency was associated much earlier with the metabolism of VFA produced in the rumen, it was not until 1956 that Armstrong, Blaxter, and coworkers reported their first study (Armstrong and Blaxter, 1956) dealing with this possibility. In experiments concerned with lipogenesis, they (Armstrong and Blaxter, 1957, 1961; Armstrong et al., 1958) infused acetic acid, propionic acid, or butyric acid (singly or in mixtures) continuously during 7-day periods into the rumen of sheep fed a maintenance ration, and the calorimetric efficiency of lipogenesis was determined by respiratory-exchange measurements and carbon-nitrogen balances determined during the final 4 or 5 days of the period. The average net efficiencies with which the ME of the VFA infused singly was utilized for fattening, were: acetic acid, 32.9%; propionic acid, 56.3; and n-butyric acid, 61.9. The ME provided by mixtures of acetic, propionic, and butyric acids in the molar proportions of 75:15:10 and 25:45:30 was utilized with a net efficiency of 31.8 and 58.1%, respectively. In similar experiments, Armstrong and Blaxter (1961) infused glucose into the rumen, abomasum, or a jugular vein and found that the net efficiencies with which the administered ME was utilized for body gain were 54.5, 71.5, and 72.8%, respectively. Thus, fermentation in the rumen reduced the net effect of glucose. When its fermentation was circumvented, the net utilization for body gain in sheep was similar to that (73.7%) observed by Nehring (1961) and Nehring et al. (1965) in rats fed glucose. Glucose is the absorbed product of starch digestion in SGA, and the net utilization for body gain of ME provided as starch to certain SGA is: chicken, 77.5% (Burlacu et al., 1967); rat, 75.5% (Nehring, 1961; Nehring et al., 1960a; Nehring et al., 1965); and pig. 75.7% (Nehring, 1961; Nehring et al., 1962; Nehring et al., 1965). In their 1965 study, Nehring et al. found a value of 64.1% for the net efficiency with which the ME, provided as starch, was utilized for body gain by sheep and cattle.

As a consequence of such observations, it appeared that the absorbed VFA resulting from the fermentation of carbohydrate in the rumen, and especially acetic acid, contributed to the high heat increment in the ruminant (Armstrong and Blaxter, 1957, 1961). Accordingly, it appeared that the difference in the absorbed products of digestion might account for the lower energetic efficiency of ruminants than of SGA. For some time it has been known that such dietary characteristics as a high ratio of concentrates to forage, and forages having a low degree of fibrousness or a small particle size, favor a ruminal fermentation in which the ratio of propionic acid to acetic acid is high. The same dietary characteristics result in a high energetic efficiency.

On the other hand, the results of a number of experiments, of which that of Bull et al. (1970) is representative, indicate that the concentration per se of acetic acid, relative to that of propionic acid in the ruminal substrate, does not influence the energetic efficiency of body gain in sheep. An explanation for the discrepancy is not apparent, unless efficiency is not improved above a certain quantity of glucogenic substances.

The carbohydrate economy of the ruminant is precarious. Under usual conditions, essentially no glucose per se is absorbed; therefore, the ruminant depends almost entirely on gluconeogenesis from noncarbohydrate substances for its glucose supply. Under certain dietary conditions, some glucose is absorbed, however. Since propionic acid provides 50 to 60% and glycerol probably somewhat less than 5% of the glucose requirement, considerable quantities of protein must be deaminated to produce the remainder. Because of the energy cost of securing glucose in this manner and of synthesizing and excreting the resultant urea (as described below), it is possible that some of the difference in the energetic efficiency between ruminants and SGA generally, and within ruminants between high-acetate- and high-propionate-producing diets, as observed in some experiments, might be explained at least in part.

Although "work of digestion" has been considerably maligned as a factor contributing to the relative inefficiency of the ruminant, present evidence is not sufficient to discard it. This viewpoint is sustained by the studies of the effects of particle size of hay on the energetics of sheep (Experiments 14 and 15, cited in Table 7-3). In the study of Blaxter and Graham (1956), a comparison of the energy losses accompanying the ingestion of coarsely chopped hay or of the same hay in finely ground form, at each of two levels of intake, indicates that the sum of the differences in the fecal and CH_4 energy losses is approximately equal to the differences in heat output. Since heat produced in the muscular work of digestion is not distinguishable from other metabolic heat, the higher output of heat by animals ingesting the chopped hay might reflect in part a greater muscular work of digestion. Whether their higher ME requirement of maintenance (Experiments 14 and 15, cited in Table 7-3) also reflects increased gastrointestinal work is not known. Although qualitative differences in the substrates (e.g., glucogenic substances), resulting from diets of different particle sizes, might be responsible for the differences in the maintenance requirements as well as in the efficiency with which

ME is utilized above maintenance, present information does not reveal them.

Relationship of metabolizability of energy and the dietary protein level to the net utilization of energy. An examination of the data obtained in many experiments conducted with ruminants suggests that usually when the ME value of the diet is high, the amount of ME required for maintenance is lower and the rate of utilization for productive functions is higher than when the ME value of the diet is low. This phenomenon was observed by van Es and Nijkamp (1967a,b) and by Moe et al. (1970) in their studies of the energy utilization by nonpregnant, nonlactating cows and by lactating cows fed a variety of diets.

In addition, they found that the maintenance requirement for ME increased as the digestible protein content of the ration was increased. Garrett (1970) reported that beef cattle fed diets containing 19 to 23% of crude protein required 20% more feed to maintain energy equilibrium than did those fed diets containing 12 to 13% of crude protein. The most extensive study made of the influence of excess protein intake on the utilization of energy by dairy cows involved the multiple regression analysis of the data obtained in 543 balance trials (Tyrrell et al., 1970). Partitioning of the ME ingested among the costs of milk production, maintenance, the products of conception, the amount of excess nitrogen ingested, and the DE concentration in the diet, revealed that the ME expense of metabolizing amino acids ingested in excess of the protein requirement is 7.2 Kcal per gram of digestible nitrogen. Although animals can utilize the carbon chains of excess amino acids as an energy source, this process is effected with an increased metabolic expense associated with the excretion of the excess nitrogen. It is of interest to note that the total cost of utilizing excess amino acids as an energy source (7.2 Kcal per gram of digestible nitrogen), as observed by Tyrrell et al. (1970), is almost twice the energy cost (3.8 Kcal per gram of nitrogen) for the mere synthesis of urea from ammonia, as estimated by Martin and Blaxter (1964).

Energetic Efficiency of Other Body Functions of Animals

Some of the conditions affecting the ME requirement of maintenance and the utilization of ME for body gain in various animals were examined. The resulting net efficiency of body energy gain in certain animals ingesting their usual range of diets was generalized as shown in Table 7-5. Also in Table 7-5 are tabulated the net efficiencies of other body functions, as generalized from various sources. The bases of some of the values are documented elsewhere in this report or in a previous one

Table 7-5. Net Efficiency of ME Utilization for Support of Certain Body Functions in Various Animals

	Net utilization of ME (%)		
Animal	*Maintenance*	*Body gain*	*Lactation or egg production*
Man	85-95	75-90	85-95
Pig	80-85	75-85	75-85
Calf (milk-fed)	90	85	—
Chicken	85-90	75-85	80-85
Rabbit	75-85	60-70	70-75
Guinea pig	75-85	60-70	70-75
Horse	75-85	60-70	70-75
Ruminants*	70-85	30-62 (70)†	40-70

* The data for ruminants represent diets ranging from good-quality forages to those high in concentrates; the net efficiency with which the ME provided by a poor-quality roughage such as wheat straw is utilized for body gain may be as low as 8%.

† The value in parentheses represents the efficiency of body gain occurring simultaneously with lactation.

(Reid, 1961); in some instances, the data employed were not derived in calorimetric experiments, but were estimated from other energy-input and animal-response data or predicted from data for animals with similar physiological characteristics.

The efficiency with which dietary ME is utilized is greater for preventing a loss of body energy (i.e., to support the maintenance function) than for other body functions. On the other hand, the gain of the conceptus is the energetically least efficient body function that has been studied. For example, in the ewe (Graham, 1964) and the cow (van Es, 1961), the net efficiency with which ME was utilized for reproduction was 13 and 25%, respectively. Moe et al. (1970) estimated that the efficiency with which ME is utilized for bovine fetal growth is 11 to 12%. The study of Brockway et al. (1963) revealing that the energy cost of maintaining pregnancy in ewes is 39.5 Mcal (heat loss in excess of that lost by nonpregnant ewes), also indicates the relatively low efficiency of reproduction. Pregnant cows required 33% more of ME during the last 60 to 90 days of gestation than did nonpregnant, nonlactating cows of the same size, according to Flatt et al. (1967).

As compared with maintenance, lactation, or egg production (by the chicken), body gain is usually less efficient. The difference appears not to be as great in SGA, however, as it is in NRH or ruminants. The data listed for ruminants in Table 7-5 represent mainly body gain by male castrates and, of course, lactation by females only. When body gain occurs simultaneously with milk production in the female ruminant, its efficiency appears to be of the same order of magnitude as that of milk production (Armstrong and Blaxter, 1965; Moe and Flatt, 1969; Moe et al., 1970). Whether this phenomenon also occurs in other species has not been determined, but it may be a characteristic only of those animals whose substrate for lipogenesis is chiefly VFA.

The estimate listed in Table 7-5 for the efficiency of lactation by woman is based chiefly on that of body gain and the relative efficiency of these two body functions in other SGA. The recent study of Thomson et al. (1970) indicated that the net conversion of dietary GE to milk energy is 96.6% as an average. Their experiment concerned 55 women examined during 7-day periods at approximately two months post-partum; 23 of them were lactating. The energy intakes were estimated from measured intakes of food. Allowances were made for the energy equivalents of body-weight changes and physical activity. Basal metabolism was estimated from height and weight tables, and the milk energy output was estimated from the body-weight changes of the suckling babies. Similar data, except for the milk energy output, were determined for 32 control women whose babies were bottle-fed. The average milk energy output (597 Kcal/day) as a percentage of the average amount of GE (618 Kcal/day) available to support lactation was 96.6. Although it seems unlikely that even the metabolizability of GE would be this high for usual diets, the value does suggest that the efficiency of lactation in woman is quite high.

Summary

The metabolism of the whole animal is influenced by a variety of dietary characteristics or treatments and by certain characteristics of the environment, and it is associated with the physiological state of the animal. Some of the influences are exerted *above* the maintenance level of

Table 7-6. Influence of Dietary Characteristics, Environmental Conditions, and Physiological States on Heat Production by Animals

Dietary characteristic or other condition	*Influence exerted*	
	Above maintenance level	*At maintenance level*
Protein deficiency	Yes	No
Protein excess	No	Yes (cattle)
Essential amino acid deficiency	Yes	(Needs study)
Fat level	Yes (rat, chicken, sheep); No (man)	No
Phosphorus deficiency	Unsettled; theoretically, yes	No
Calcium deficiency	Undetermined	Probably; basal metabolic rate elevated
Magnesium deficiency	No (rat, calf)	Yes
Sodium deficiency	Yes	Undetermined
Iron and copper deficiency	Yes	Undetermined
Vitamin A deficiency	No (rat, calf)	No (rat, calf)
Vitamin D deficiency	No	Probably; basal metabolic rate elevated
B-complex vitamin deficiency*	Yes (needs study)	Undetermined
Physical form of diet	Yes (ruminants)	Yes (ruminants)
Meal frequency	Yes (some animals†)	Probably; unsettled‡
Physical activity	No	Yes
Sex	Yes (sheep)	No (sheep)
Compensatory growth	Yes	(Needs study)
Body fatness	No (sheep, cattle)	Unsettled§
Ambient temperature	No#	Yes
Altitude	No (rat)	Yes (rat)

* Only thiamine, riboflavin, pyridoxine, pantothenate, folacin, and Vitamin B_{12} appear to have been studied. Since only one level of each diet, deficient and adequate, was fed, it is not possible to determine whether a deficiency affected the maintenance requirement; however, the energy-gain response was so marked that the influence apparently is exerted at least above the maintenance level.

† In rats, a greater energy storage results from infrequent meals than from frequent meals providing the same daily intake of energy; in ruminants the opposite effect occurs.

‡ Whether the frequency of meals of isocaloric diets affects the energy requirement of maintenance has not been determined, but in rats it is known that meal eaters are less active physically than are those which are allowed to nibble frequently.

§ The most detailed experiment conducted with sheep indicated that the ME requirement of maintenance is not affected by body fatness (Graham, 1969); a more limited experiment with cattle suggested a higher maintenance cost for fat animals than for thin ones (Armsby and Fries, 1917).

\# An exception to this conclusion might occur at a high temperature in fat animals ingesting a high level of energy; however, it is unlikely that such would occur because a high ambient temperature is incompatible with a high intake of food.

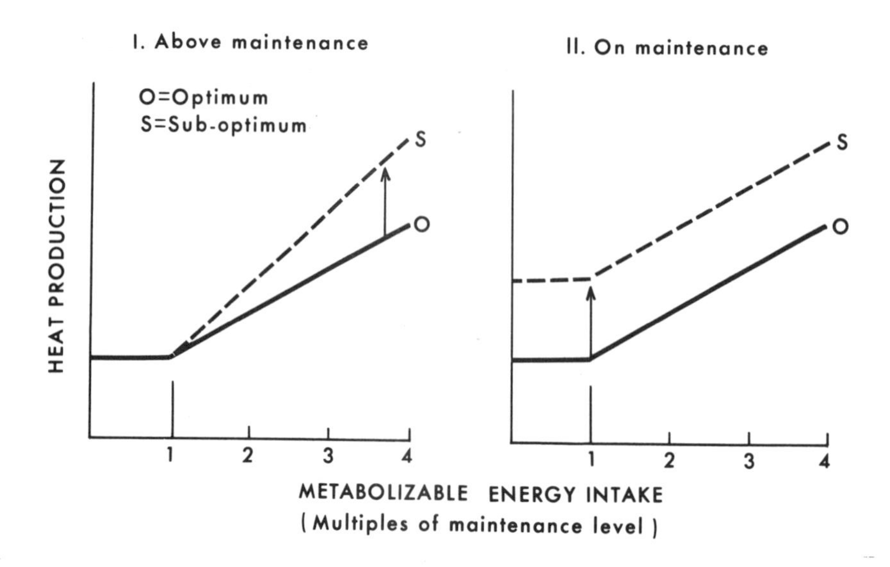

Figure 7-2. Effects on metabolism exerted by dietary characteristics and the environment or associated with physiological states of the animal.

energy input, and others are exerted *below* or at the maintenance level, but the effect so generated is superimposed more or less uniformly on the metabolic rates occurring above the maintenance level. In addition, certain dietary influences are exerted both above and below the maintenance level. When the influence is exerted only above the maintenance level situation I in Figure 7-2 exists; when the influence is exerted at the maintenance level, situation II in Figure 7-2 exists. Figure 7-2 also represents a qualitative diagram of the direction of the effect on heat production when the situation created by various influences (itemized in Table 7-6) is at optimum (O) (e.g., when the diet is adequate, the diet is not deficient in a required nutrient, or the ambient temperature is within the range of thermal neutrality), or suboptimum (S) (e.g., when the diet is inadequate, or deficient in a required nutrient, or the ambient temperature is below or above the range of thermal neutrality). Thus, the generalizations listed in Table 7-6 and described in Figure 7-2 represent a summary of the effects of some dietary characteristics, environmental conditions, and animal states on heat production or, conversely, on the energetic efficiency of the whole animal. It will be noted that some of the appraisals are conditional or qualified, and that the effects of many dietary, environmental, or animal characteristics are yet to be studied.

Some of the studies reported here were supported by Grant AM-02889 from the National Institute of Arthritis and Metabolic Diseases, U.S. Public Health Service.

References

Andersen, P. E., Reid, J. T., Anderson, M. J., and Stroud, J. W., 1959. Influence of level of intake upon the apparent digestibility of forages and mixed diets by ruminants. *J. Anim. Sci.* 18:1299.

Armsby, H. P., and Fries, J. A., 1917. Influence of the degree of fatness of cattle upon their utilization of feed. *J. Agr. Res.* 11:451.

Armstrong, D. G., and Blaxter, K. L., 1956. The heat increment of a mixture of steam-volatile fatty acids and of glucose in the fasting sheep. *Proc. Brit. Nutr. Soc.* 15:i.

Armstrong, D. G., and Blaxter, K. L., 1957. The utilization of acetic, propionic, and butyric acids by fattening sheep. *Brit. J. Nutr.* 11:413.

Armstrong, D. G., and Blaxter, K. L., 1961. The utilization of the energy of carbohydrate by ruminants. *Europ. Assoc. Anim. Prod. Publ.* 10:187.

Armstrong, D. G., and Blaxter, K. L., 1965. Effects of acetic and propionic acids on energy retention and milk secretion in goats. *Europ. Assoc. Anim. Prod. Publ.* 11:59.

Armstrong, D. G., Blaxter, K. L., Graham, N. McC., and Wainman, F. W., 1958. The utilization of the energy of two mixtures of steam-volatile fatty acids by fattening sheep. *Brit. J. Nutr.* 12:177.

Behnke, A. R., 1953. The relation of lean body weight to metabolism and some consequent systematizations. *Ann. N. Y. Acad. Sci.* 56:1095.

Black, A., Maddy, K. H., and Swift, R. W., 1950. The influence of low levels of protein on heat production. *J. Nutr.* 42:415.

Blaxter, K. L., 1952. The nutrition of the young Ayrshire calf. 6. The utilization of the energy of whole milk. *Brit. J. Nutr.* 6:12.

Blaxter, K. L., 1962. *The Energy Metabolism of Ruminants.* Charles C Thomas, Springfield, Ill., p. 275.

Blaxter, K. L., and Graham, N. McC., 1955. Plane of nutrition and starch equivalents. *J. Agr. Sci.* 46:292.

Blaxter, K. L., and Graham, N. McC., 1956. The effect of the grinding and cubing process on the utilization of the energy of dried grass. *J. Agr. Sci.* 47:207.

Blaxter, K. L., and Wainman, F. W., 1961. The utilization of food by sheep and cattle. *J. Agr. Sci.* 57:419.

Blaxter, K. L., and Wainman, F. W., 1964. The utilization of the energy of different rations by sheep and cattle for maintenance and for fattening. *J. Agr. Sci.* 63:113.

Blaxter, K. L., and Wainman, F. W., 1966. The fasting metabolism of cattle. *Brit. J. Nutr.* 20:103.

Breirem, K., 1939. Der Energieumsatz bei den Schweinen. *Tierernährung,* 11:487.

Brockway, J. M., McDonald, J. D., and Pullar, J. D., 1963. The energy cost of reproduction in sheep. *J. Physiol.* 167:318.

Brody, S., and Procter, C. R., 1932. Growth and development with special reference to domestic animals. XXIII. Relation between basal metabolism and mature body weight in different species of mammals and birds. *Mo. Agr. Exp. Sta., Res. Bull.* 166.

Bull, L. S., Reid, J. T., and Johnson, D. E., 1970. Energetics of sheep concerned with the utilization of acetic acid. *J. Nutr.* 100:262.

Burlacu, G., Grossu, D., Marinescu, G., Baltac, M., and Grunca, D., 1967. Efficiency of utilization of the energy of starch in birds. *Europ. Assoc. Anim. Prod. Publ.* 12:369.

Buskirk, E. R., Thompson, R. H., Lutwak, L., and Whedon, G. D., 1963. Energy balance of obese patients during weight reduction: influence of diet restriction and exercise. *Ann. N. Y. Acad. Sci.* 110:918.

Cairnie, A. B., and Pullar, J. D., 1957. The metabolism of the young pig. *J. Physiol.* 139:15P.

Chinn, K. S. K., and Hannon, J. P., 1969. Efficiency of food utilization at high altitude. *Fed. Proc.* 28:944.

Chudy, A., and Schiemann, R., 1967. Utilization of dietary fat for maintenance and fat deposition in model studies with rats. *Europ. Assoc. Anim. Prod.* 12:161.

Cook, C. W., Stoddart, L. A., and Harris, L. E., 1952. Determining the digestibility and metabolizable energy of winter range plants by sheep. *J. Anim. Sci.* 11:578.

Cowan, R. L., Long T. A., and Jarrett M., 1968. Digestive capacity of the meadow vole *(Microtus pennsylvanicus). J. Anim. Sci.* 27:1517.

Crawford, A., 1788. *Experiments and Observations on Animal Heat and the Inflammation of Combustible Bodies* (ed. 2). London.

Cuthbertson, D. P., and Chalmers, M. I., 1950. Utilization of a casein supplement administered to ewes by ruminal and duodenal fistulae. *Biochem. J.* 46:xvii.

Dawson, T. J., and Hulbert, A. J., 1970. Standard metabolism, body temperature, and surface areas of Australian marsupials. *Am. J. Physiol.* 218:1233.

Elsden, S. R., Hitchcock, M. W. S., Marshall, R. A., and Phillipson, A. T., 1946. Volatile acids in the digesta of ruminants and other animals. *J. Exp. Biol.* 22:191.

Es., A. J. H. van, 1961. Between-animal variation in the amount of energy required for the maintenance of cows. Ph.D. diss. Wageningen, The Netherlands.

Es, A. J. H. van, Brouwer, E., and Nijkamp, H. J., 1961. Within-animal variation in maintenance requirement of metabolizable energy in cows. *Europ. Assoc. Anim. Prod. Publ.* 10:131.

Es, A. J. H. van, Brouwer, E., and Nijkamp, H. J., 1965. The feeding value for the maintenance of cows of hay from six parts of the Netherlands. *Europ. Assoc. Anim. Prod. Publ.* 11:95.

Es, A. J. H. van, and Nijkamp, H. J., 1967a. Energy, carbon and nitrogen balance experiments with non-lactating, non-pregnant cows. *Europ. Assoc. Anim. Prod. Publ.* 12:203.

Es, A. J. H. van, and Nijkamp, H. J., 1967b. Energy, carbon and nitrogen balance experiments with lactating cows. *Europ. Assoc. Anim. Prod. Publ.* 12:209.

Evans, G. H., 1910. *Elephants and Their Diseases.* Government Printing Office, Burma (Rangoon).

Fernandez-Baca, S., 1966. Utilización comparativa de los forrajes por la Alpaca y el Ovino. *Proc. 5th PanAmer. Congr. Vet. Med. and Zootech.* 1:352.

Flatt, W. P., Moe, P. W., and Moore, L. A., 1967. Influence of pregnancy and ration composition on energy utilization by dairy cows. *Europ. Assoc. Anim. Prod. Publ.* 12:123.

Forbes, E. B., Black, A., Thacker, E. J., and Swift, R. W., 1939. The utilization of energy-producing nutriment and protein as affected by the level of the intake of beef muscle protein. *J. Nutr.* 18:47.

Forbes, E. B., Braman, W. W., Kriss, M., Jeffries, C. D., Swift, R. W., French, R. B., Miller, R. C., and Smythe, C. V., 1928. The energy metabolism of cattle in relation to the plane of nutrition. *J. Agr. Res.* 37:253.

Forbes, E. B., Braman, W. W., Kriss, M., Swift, R. W., French, R. B., Smythe, C. V., Williams, P. S., and Williams, H. H., 1930. Further studies of the energy metabolism of cattle in relation to the plane of nutrition. *J. Agr. Res.* 40:37.

Forbes, E. B., Swift, R. W., Bratzler, J. W., Black, A., Wainio, W. W., Marcy, L. F., Thacker, E. J., and French, C. E., 1941. The minimum base value of heat production in animals: A research in the energy metabolism of cattle. *Pa. Agr. Exp. Sta. Bull.* 415.

Forbes, E. B., Swift, R. W., Thacker, E. J., Smith, V. F., and French, C. E., 1946. Further experiments on the relationship of fat to economy of food utilization. II. By the mature albino rat. *J. Nutr.* 32:397.

Garrett, W. N., 1970. Energy utilization of high protein diets. *J. Anim. Sci.* 31:242.

Graham, N. McC., 1964. Energy exchanges of pregnant and lactating ewes. *Austral. J. Agr. Res.* 15:127.

Graham, N. McC., 1966. The metabolic rate of fasting sheep in relation to total and lean body weight, and the estimation of maintenance requirements. *Austral. J. Agr. Res.* 18:127.

Graham, N. McC., 1969. The influence of body weight (fatness) on the energetic efficiency of adult sheep. *Austral. J. Agr. Res.* 20:375.

Golley, F. B., 1960. Anatomy of the digestive tract of *Microtus. J. Mamm.* 41:89.

Hamilton, T. S., 1939. The growth, activity and composition of rats fed diets balanced and unbalanced with respect to protein. *J. Nutr.* 17:565.

Hellberg, A., 1949. *Metabolism of Rabbits at Different Planes of Nutrition.* Almquist and Wiksells Boktryckeri, Uppsala.

Henning, S. J., and Bird, F. J. R., 1970. Concentrations and metabolism of volatile fatty acids in the fermentative organs of two species of kangaroo and the guinea pig. *Brit. J. Nutr.* 24:145.

Hintz, H. F., 1969. Effect of coprophagy on digestion and mineral excretion in the guinea pig. *J. Nutr.* 99:375.

Hintz, H. F., Lowe, J. E. and Schryver, H. F., 1969. Protein sources for horses. *Proc. Cornell Nutr. Conf.,* p. 65.

Hoover, W. H., Mannings, C. L., and Sheerin, H. E., 1968. Observations pertaining to digestion in the golden hamster. *J. Anim. Sci.* 27:1512.

Johnson, W. L., 1966. The nutritive value of *Panicum Maximum* (Guinea Grass) for cattle and water buffaloes in the tropics. Ph.D. diss. Cornell University.

Kellner, O., 1905. *Die Ernährung der Landwirtschaftlichen Nutzliere* (ed. 9). Paul Parey, Berlin.

Keys, J. E., Jr., and Van Soest, P. J., 1970. Digestibility of forages by the meadow vole *(Microtus pennsylvanicus). J. Dairy Sci.* 53:1502.

Kielanowski, J., 1965. Estimates of the energy cost of protein deposition in growing animals. *Europ. Assoc. Anim. Prod. Publ.* 11:13.

Knudsen, K. A., 1968. Energy metabolism and body composition of growing male guinea pigs fed at three levels of energy intake. Master's Thesis. Cornell University.

Lavoisier, A. L., 1777. Expériences sur la respiration des animaux et sur les changements qui arrivant a l'air en passant par leur poumons. *Mém. de l'Acad. des Sci.* p. 185.

Lavoisier, A. L., and LaPlace, P. S., 1780. Mémoire sur la chaleur. *Mém. de l'Acad. des Sci.* p. 379.

Lee, C., and Horvath, D. J., 1968. Management of meadow vole *(Microtus pennsylvanicus). J. Anim. Sci.* 27:1517.

Lee, R. C., 1942. Heat production of the rabbit at 28 °C as affected by previous adaptation to temperatures between 10° and 31 °C. *J. Nutr.* 23:83.

Marston, H. R., 1948. Energy transactions in the sheep. *Austral. J. Sci. Res.* 1:93.

Martin, A. K., and Blaxter, K. L., 1964. The energy cost of urea synthesis in sheep. *Europ. Assoc. Anim. Prod. Publ.* 11:83.

Matsumoto, T., 1955. Nutritive value of urea as a substitute for feed protein. 1. Utilization of urea by the golden hamster. *Tohoku J. Agr. Res.* 6:127.

Mering, J. von, and Zuntz, N., 1877. In wiefern beeinflusst Nährungszuführ die thierischen Oxydationsprocesse? *Pfluegers Arch. Physiol. Menschen Tiere* 15:634.

Miller, A. T., and Blyth, C. S., 1953. Estimation of lean body mass and body fat from basal oxygen consumption and creatinine excretion. *J. Appl. Physiol.* 5:73.

Mitchell, H. H., Hamilton, T. S., McClure, F. J., Haines, W. T., Beadles, J. R., and Morris, H. P., 1932. The effect of the amount of feed consumed by cattle on the utilization of its energy content. *J. Agr. Res.* 45:163.

Moe, P. W., and Flatt, W. P., 1969. Use of body tissue reserves for milk production by the dairy cow. *J. Dairy Sci.* 52:928.

Moe, P. W., Tyrrell, H. F., and Flatt, W. P., 1970. Partial efficiency of energy use for maintenance, lactation, body gain, and gestation in the dairy cow. *Europ. Assoc. Anim. Prod. Publ.* 13:65.

Moir, R. J., 1965. The comparative physiology of ruminant-like animals. In *Physiology of Digestion in the Ruminant* (R. W. Dougherty, R. S. Allen, W. Burroughs, N. L. Jacobson, and A. D. McGilliard, eds.). Butterworths, Washington, p. 1.

Moir, R. J., Somers, M., and Waring, H., 1956. Studies on marsupial nutrition. 1. Ruminant-like digestion in a herbivorous marsupial *(Setonix brachyurus,* Quoy and Gaimard). *Austral. J. Biol. Sci.* 9:293.

Nehring, K., 1958. Aufgaben und Arbeiten des Oskar Kellner-Institutes für Tierernährung, Rostock, auf dem Gebiet der Respirationsversuche. *Europ. Assoc. Anim. Prod. Publ.* 8:248.

Nehring, K., 1961. Probleme der einheitlichen Bewerlung der Futterstoffe auf der Grundlage der Netto-Energie. *Europ. Assoc. Anim. Prod. Publ.* 10:307.

Nehring, K., Jentsch, W., and Schiemann, R., 1960a. Die Verwertung reiner Nährstoffe. 1. Mitt.:Versuche mit Kaninchen und Ratten. *Arch. Tierernähr.* 11:233.

Nehring, K., Schiemann, R., Hoffman, L., and Klippel, W., 1960b. Die Verwertung der Futterenergie in Abhängigkeit vom Ernährungsniveau. 2. Mitt. Versuche mit Schweinen. *Arch. Tierernähr.* 10:275.

Nehring, K., Schiemann, R., Hoffman, L., and Jentsch, W., 1962. Die Verwertung reiner Nährstoffe. 4. Mitt.: Vergleich der Tierarten. *Arch. Tierernähr.* 11:359.

Nehring, K., Schiemann, R., Hoffman, L., Klippel, W., and Jentsch, W., 1965. Utilization of the energy of cellulose and sucrose by cattle, sheep, and pigs. *Europ. Assoc. Anim. Prod. Publ.* 11:249.

Ørskov, E. R., Fraser, C., and Corse, E. L., 1970. The effect on protein utilization of different protein supplements via the rumen or via the abomasum in young growing sheep. *Brit. J. Nutr.* 24:803.

Paladines, O. L., Reid, J. T., Van Niekerk, B. D. H., and Bensadoun, A., 1964. Energy utilization by sheep as influenced by the physical form, composition and level of intake of diet. *J. Nutr.* 83:49.

Parker, J. W., and Clawson, A. J., 1967. Influence of level of total feed intake on digestibility, rate of passage, and energetic efficiency of reproduction in swine. *J. Anim. Sci.* 26:485.

Passmore, R., Meiklejohn, A. P., Dewar, A. D., and Thow, R. K., 1955. Energy utilization in overfed thin young men. *Brit. J. Nutr.* 9:20.

Petrie, J. G., 1970. The Effects of Cage Density and Simulated Cover on Food Utilization and Other Processes in the Meadow Vole *(Microtus pennsylvanicus).* Master's thesis. Cornell University.

Reid, J. T., 1961. Nutrition of lactating farm animals. In *Milk: the Mammary Gland and Its Secretion* (S. K. Kon and A. T. Cowie, eds.). Academic Press, New York, p. 47.

Reid, J. T., and Tyrrell, H. F., 1964. Effect of level of intake on the energetic efficiency of animals. *Proc. Cornell Nutr. Conf.,* p. 25.

Rogerson, A., 1968. Energy utilization by the eland and wildebeest. In *Comparative Nutrition of Wild Animals* (M. A. Crawford, ed.). Acadmic Press, New York, p. 153.

Schiemann, R., 1963. Die energetische Verwertung des Proteins. *Deutsche Akad. Landw., Sitzungsberichte* 12:39.

Schürch, A., 1961. Die energetische Wirkung verschieden hoher Eiweiss-und Stärkegaben, gemessen am wachsenden und ausgewachsenen Schaf. *Europ. Assoc. Anim. Prod. Publ.* 10:250.

Shannon, D. W. F., and Brown, W. O., 1969. Calorimetric studies on the effect of dietary energy source and environmental temperature on the metabolic efficiency of energy utilization by mature Light Sussex cockerels. *J. Agr. Sci.* 72:479.

Slade, L., and Robinson, D. W., 1969. Utilization of protein and NPN by adult mares. *J. Anim. Sci.* 29:144.

Swift, R. W., Barron, G. P., Fisher, K. H., Cowan, R. L., Hartsook, E. W., Hershberger, T. V., Keck, E., King, R. P., Long, T. A., and Berry, M. E., 1959. The utilization of dietary protein and energy as affected by fat and carbohydrate. *J. Nutr.* 68:281.

Swift, R. W., Barron, G. P., Jr., Fisher, K. H., Magruder, N. D., Black, A., Bratzler, J. W., French, C. E., Hartsook, E. W., Hershberger, T. V., Keck, E., and Stiles, F. P., 1957. Relative dynamic effects of high versus low protein diets of equicaloric content. *Pa. Agr. Exp. Sta. Bull.* 618.

Swift, R. W., Bratzler, J. W., James, W. H., Tillman, A. D., and Meek, D. C., 1948. The effect of dietary fat on utilization of the energy and protein of rations by sheep. *J. Anim. Sci.* 7:475.

Tadayyon, B., and Lutwak, L., 1969. Role of coprophagy in utilization of triglycerides, calcium, magnesium, and phosphorus in the rat. *J. Nutr.* 97:243.

Thacker, E. J., and Brandt, C., 1955. Coprophagy in the rabbit. *J. Nutr.* 55:375.

Thomas, O. P., 1966. Studies on lysine for chickens. *Proc. Maryland Nutr. Conf.,* p. 80.

Thomson, A. M., Hytten, F. E., and Billewicz, W. Z., 1970. The energy cost of human lactation. *Brit. J. Nutr.* 24:565.

Thorbek, G., 1967. Studies of the energy metabolism of growing pigs. *Europ. Assoc. Anim. Prod. Publ.* 12:281.

Tomme, M., and Missiutkina, M., 1936. Der Einfluss der Lufttemperatur auf den Gas-und Stoffumsatz beim Kaninchen. *Tierernährung* 8:97.

Tyrrell, H. F., Moe, P. W., and Flatt, W. P., 1970. Influence of excess protein intake on energy metabolism of the dairy cow. *Europ. Assoc. Anim. Prod. Publ.* 13:69.

Vercoe, J. E., 1965. The effect of the level of dietary protein on the nitrogen and energy metabolism of adult guinea pigs. *Europ. Assoc. Anim. Prod. Publ.* 11:285.

Vercoe, J. E., 1970. The fasting metabolism of Brahman, Africander and Hereford x Shorthorn cattle. *Brit. J. Nutr.* 24:599.

8

Control of Thermoregulatory Behavior

John D. Corbit III

This chapter presents an analysis of the neural mechanism responsible for the control of thermoregulatory behavior (Corbit, 1969, 1970, 1973; Corbit and Ernits, 1974), and shows how this behavioral mechanism contributes to the regulation of internal body temperature. First an overview of the general problem of the behavioral contribution to homeostasis is briefly outlined; this conceptual framework helped to motivate the experiments described under Analysis.

Behavioral Contribution to Homeostasis

Maintenance of the relative constancies of many aspects of the internal environment depends on behavioral interactions with the external environment as well as on automatic adjustments within the body. For example, nutrient, fluid, and thermal balance can not be maintained in the long run without appropriate control of the behavioral activities of eating, drinking, and selection of a favorable thermal environment. In each case, these behavioral activities contribute to homeostasis by adjusting the rate of exchange of the regulated quantity (nutrients, fluid, or heat) between the external environment and the interior of the body in such a way as to favor the constancy of the body's content of each quantity.

Generalized Control System

From the point of view of control theory (Grodins, 1963; Milhorn, 1966; Riggs, 1963, 1970; Yamamoto and Brobeck, 1965), regulatory behavior is regarded as the controller action in a negative feedback control system involving the nervous system and its environment. The systems controlling eating, drinking, and thermoregulatory behavior are regulators of the body's content of nutrients, fluid, and heat.

Shown below is a block diagram model of a generalized behavioral control system (Model 1). The blocks represent structural components of the system and the arrows indicate the direction of the flow of information between components. The components of this generalized control system are specified in highly abstract and general terms. As a result, the model is able to capture the essence of the problem, and to call attention to the fundamental properties common to all behavioral regulatory systems. The specific systems responsible for the control of each of the regulatory behaviors mentioned above are all assumed to be of this general type.

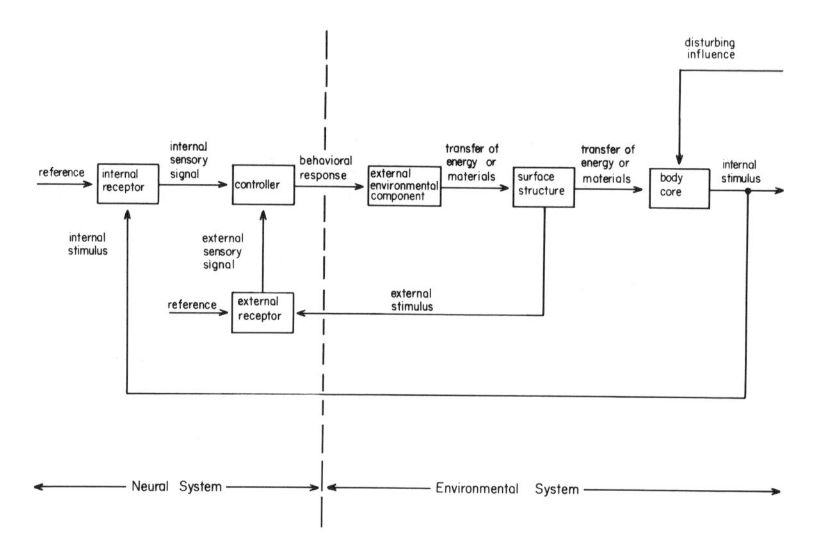

At the most general level of organization, the behavioral control system in Model 1 consists of two subsystems, an active or controlling (neural) system, which is of primary interest here, and a passive or controlled (environmental) system. The boundary or interface between the neural system (which includes the nervous system and its effectors) and the environment of the neural system (which includes the internal and external environments of the animal) is indicated by the vertical broken line. The neural system converts changes in internal and external stimuli on the input side to changes in behavioral responses on the output side. The environmental system converts changes in behavioral response to changes in internal and external stimuli.

At the next level of organization, the neural and environmental subsystems have each been reduced to a set of three components (blocks). The neural system consists of internal and external receptors which convert internal and external stimuli to sensory signals, and a controller (neural network plus exterofectors) which converts sensory signals to behavioral responses. The environmental system consists of an external environmental component and surface and core components of the body. The components of the system are arranged so as to form two closed loops, one involving the internal stimulus and the other involving the external stimulus. The function of the control system as a regulator can be seen by tracing the sequence of events as information flows around the two loops.

Function of the Generalized Control System

Internal stimulus loop. The loop involving the internal stimulus contributes to the maintenance of the constancy of the internal environment in the following way. The internal stimulus (e.g., blood glucose concentration, intracellular fluid volume, or internal body temperature) varies with and is thus a measure of the amount of the regulated quantity (e.g., nutrients, fluid, or heat) contained in the body core. When a disturbing influence causes the content of the body core to deviate from its optimal value, the associated change in the value of the internal stimulus (e.g., hypoglycemia, cellular dehydration, or hypothermia) is detected by appropriately sensitive internal receptors (e.g., chemically, osmotically, or thermally sensitive interoceptors) located in the brain and elsewhere in the body core. The internal receptors convert the internal stimulus to an internal sensory signal, and the controller converts the sensory signal to a behavioral response (e.g., eating, drinking, or thermoregulatory response). The response operates on an external environmental component (a source of the regulated quantity), causing a transfer of the regulated quantity (e.g., nutrients, fluid, or heat) from the external environment to the surface structure of the body (alimentary canal or skin). Finally, the regulated quantity is transferred from the surface structure to the core, repairing the initial disturbance to its content. When the deviation of content from its optimal value has been eliminated, the system becomes quiescent.

Thus, the internal stimulus loop involves detection of an internal disturbance and activation of behavior resulting in an exchange with the external environment such that the internal disturbance is opposed or eliminated. However, there are more or less lengthy time lags in the transfer between the surface and the core of the body (e.g., times associated with the digestion, absorption, and distribution of nutrients, with the absorption and distribution of fluids, or with heat transfer between the surface and the core). These lags or delays introduced by the mechanisms of transfer between surface and core would make the system unstable, if the internal stimulus loop were the only source of control. For example, if a response (e.g., eating, drinking, or surface warming), initiated by a disturbance to the content of the core (e.g., hypoglycemia, cellular dehydration, or hypothermia), continued until that disturbance was eliminated, the lags introduced by surface-to-core transfer times guarantee that serious overshooting (e.g., overeating, overdrinking, or perhipheral overheating) would occur. Overshooting or overcommitment of this type, and the high-amplitude oscillations of the core content that would follow, are prevented by the damping action of the external stimulus loop.

External stimulus loop. When regulatory behavior is activated by an internal disturbance, the most immediate consequences of the behavior are events at the body surface. The behavioral response (e.g., eating, drinking, or thermoregulatory response) causes a change in the surface structure's content of the regulated quantity (e.g., nutrient or fluid content of the alimentary canal, or heat content of the skin). Some aspect of the content of the surface structure (e.g., the chemical or osmotic concentration or the mechanical force applied to the wall of the gastrointestinal tract, or the temperature of the skin) is a stimulus for an external receptor located in the surface structure (e.g., chemically, osmotically, or mechanically sensitive receptors in the alimentary canal, or thermally sensitive receptors in the skin). The external receptor converts the external stimulus to an external sensory signal which acts on the controller to inhibit the behavioral response. As the behavioral activity proceeds, the content of the surface structure continues to change and the external sensory signal becomes progressively more intense. At some point, the inhibitory effects of the external sensory signal become sufficient to balance the excitation arising from the internal disturbance, and the behavioral activity ceases, even though the internal disturbance still exists. In effect, the external sensory signal provides the central controller with information which is predictive of the ultimate elimination of the internal disturbance; by limiting the behavior before the internal disturbance has actually been eliminated, it serves to damp the system, preventing overshooting and oscillation of the content of the body core. Finally, with the ultimate transfer of the regulated quantity from the surface to the core, the bases for the excitation and inhibition of internal and external origin disappear and, in the end, the neural system becomes quiescent, having restored the initially disturbed content of the body core to its optimal value.

Analysis of the Neural System

In general, analysis of a system consists of identifying its components and specifying the input-output relationship for each component. At the most general level of analysis, the behavioral system consists of two components, a neural system (the path through the nervous system from receptors to effectors which converts stimulus input to response output) and an environmental system (the path through the external environment and body which converts changes in effector response to changes in stimulus input to the receptors). The problems at this level of analysis are first to define the boundary between the neural and environmental systems (i.e., to identify the adequate stimuli for the receptors and the actions of the effectors), and then to specify the rules for the conversion of changes in stimuli to changes in responses by the neural system, and for the conversion of changes in responses to changes in stimuli by the environmental system.

To understand the performance of the behavioral control system as a whole, the input-output relationships for both the neural and the environmental subsystems must be known. However, it is possible to study either subsystem in isolation. From the point of view of neurophysiology, the neural system is of greatest interest. The task is to define the information-processing operations performed by the neural system, and to explain them in terms of the underlying neural mechanism. Thus the remainder of this chapter focuses on the neural system, and greatly de-emphasizes the details of the environmental system.

Conceived of as an information-processing problem, the analysis of a given neural system must logically proceed according to the following steps. At the most general level of analysis, the first steps are to identify the inputs (adequate stimuli) and outputs (effector actions—i.e., behavioral responses), and then to specify the input-output relationship (stimulus-response relationship), treating the neural system as a whole as a "black box." The stimulus-response relationship, determined from behavioral studies of the intact neural system, defines the problem for neurophysiological investigation: that is, to explain the information-processing characteristics of the system as a whole in terms of the information-processing characteristics of the individual neural elements of the system.

The next step is to determine the information-processing characteristics of the first-order elements in the neural system. This step involves electrophysiological studies of the process of transduction performed by the receptors—i.e., the conversion of changes in adequate stimulus energy to changes in the frequency of nerve impulses. The problem is reduced to a new level of analysis by the specification of the input-output relationships for the receptors (the problem of regulatory behavior is stated at this level by block diagram Model 1 of the neural system. At this level of analysis, the rest of the neural system (second-order through *nth*-order neurons) is the "black box" (labeled "controller" in Model 1), and its input-output relationship can be deduced from knowledge of the input-output relations for the receptors and for the neural system as a whole.

The subsequent steps in the analysis of the neural system involve proceeding, neuron by neuron, with electrophysiological studies of the input-output characteristics of the second-order through *n*th-order neurons. The endpoint of the electrophysiological investigation would presumably be reduction of the neural system to the level of a neural network model, based on knowledge of the intput-output relationships for all of the neural elements making up the path from receptors to effectors.

Analysis of the Behavioral Thermoregulatory System

The Problem

At the most general level of analysis, the first step is to identify the internal and external stimuli. Important progress has already been made in this regard: hypothalamic temperature has been shown to be an internal thermal stimulus of major behavioral significance (Satinoff, 1964), and the external stimulus is skin temperature (Hardy, 1954; Weiss and Laties, 1961; Winslow et al., 1937).

After identifying the stimuli, the next step is to specify the input-output characteristics of the neural system as a whole, that is, to specify the nature (linear, logarithmic, etc.) of the independent effect of each stimulus on the behavioral response, and the nature (additive, multiplicative, etc.) of the interaction of the stimuli when both are varied. One objective of the experiments described below was to obtain quantitative statements for the independent and combined effects of hypothalamic and skin temperatures on the behavioral thermoregulatory responses of the animal.

Knowledge of the input-output characteristics of the hypothalamic (Hardy, 1969; Hellon, 1970) and cutaneous (Hensel, 1970; Iggo, 1970; Zotterman, 1959) thermal receptors makes it possible to reduce the problem from the single-component level of the neural system as a whole to the three-component level of internal and external receptors plus a controller (see diagram above). A further objective was to learn more about the information-processing operations performed by the controller for thermoregulatory behavior, inferring these operations from the results of appropriately designed behavioral experiments and knowledge of the receptor characteristics. Tentative answers were obtained to the questions of (1) how the controller combines hypothalamic and cutaneous sensory information to determine the intensity of the behavioral response, (2) how the controller uses the pattern of sensory inputs to select which particular response will be activated at a given moment, and (3) what variables are regulated by the controller.

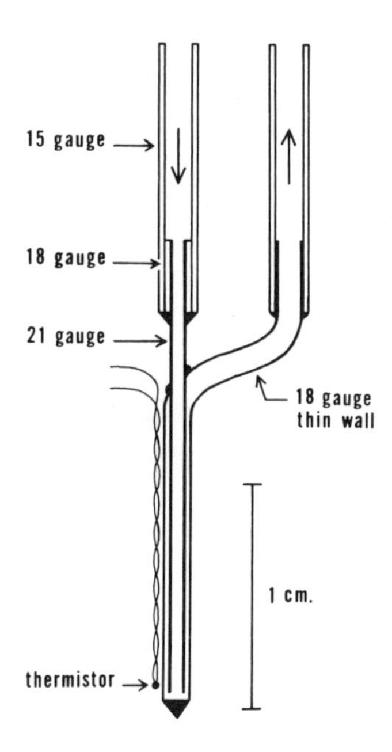

Figure 8-1. Detail of the stainless steel thermode. Arrows show direction of water flow through the thermode. A tiny thermistor attached to the thermode near its tip was used to measure hypothalamic temperature. (Corbit, 1973)

Method

Subjects. The subjects were adult male Sprague-Dawley rats (500-700 g). Thermodes, consisting of a concentric arrangement of stainless steel tubes (Figure 8-1), were permanently implanted, on the basis of sterotaxic coordinates which placed the thermode 0.5 mm or less from the anatomical midline and in contact with the rostral aspects of the medial preoptic nuclei.

Figure 8-2 shows the thermode in place in a preparation used for long-term measurement of thermal gradients in the brain. Several animals were prepared in this way, with thermistor probes at different distances from the thermode surface. Measurements of the spread of thermal effect in waking animals showed that the effects of changes in thermode temperature were minimal in the posterior hypothalamus and absent in the midbrain, and that hypothalamic temperature, as measured by the small (0.3-mm) thermistor attached to the thermode surface (Figure 8-1) is a

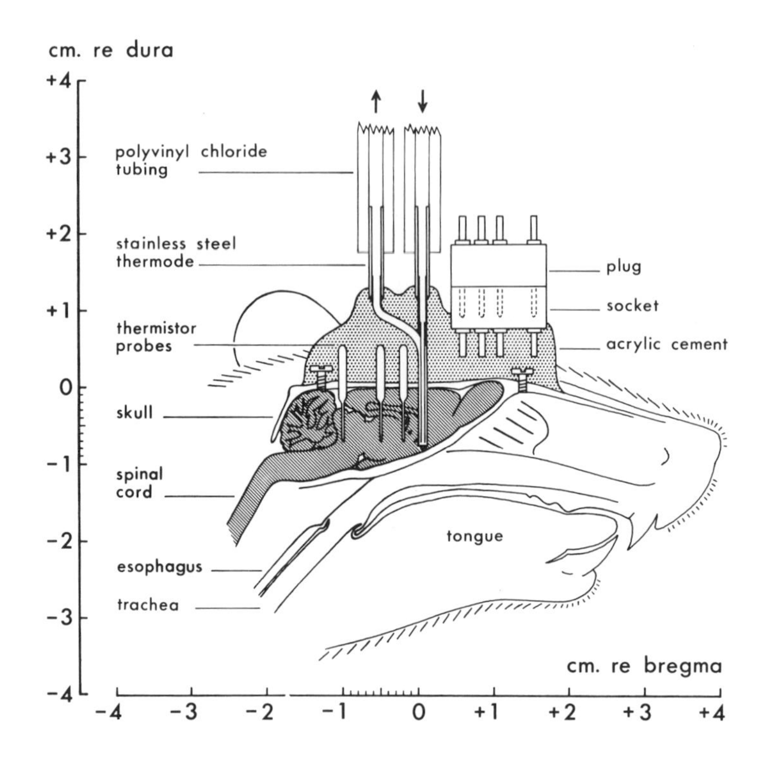

Figure 8-2. Diagram of midsagittal section through the head of a rat prepared with a preoptic thermode and with thermistor probes at 2, 5, and 10 mm posterior to the thermode surface. The wires from the thermistors to the miniature socket cemented to the skull are not shown. (Corbit, 1973)

good measure of the tissue temperature of the temperature-sensitive preoptic area.

After the animals had recovered from surgery, the effectiveness of the hypothalamic stimulus and the integrity of the preoptic area were confirmed by showing that small increases in thermode temperature elicited vasodilation of the skin of the paws and tail. Small (0.3-mm) thermistors of the type used to measure hypothalamic temperature were also used to measure air, skin, and rectal temperatures. Measurements of skin temperature were usually made on the dorsal surface of the tail about 4 cm from its base.

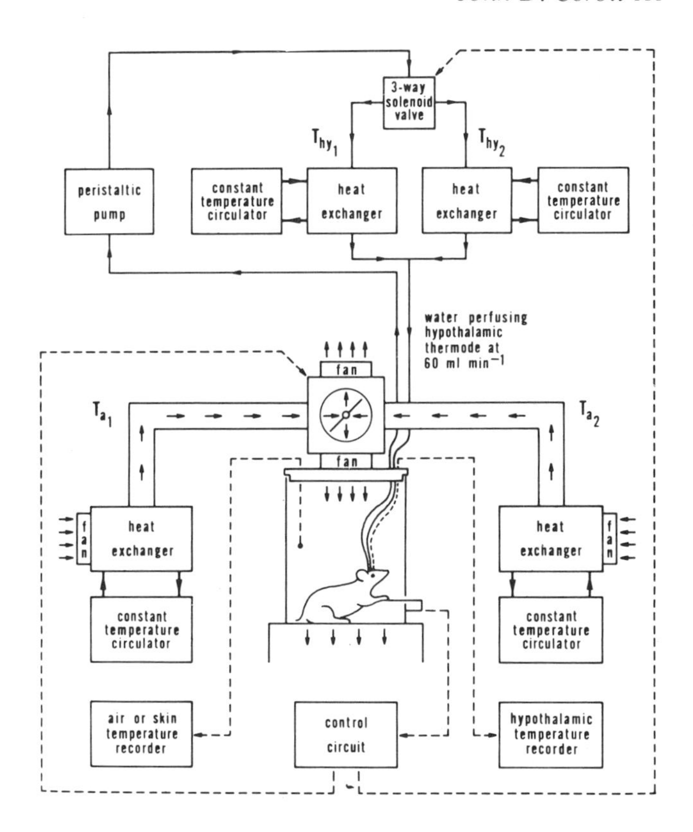

Figure 8-3. System for converting the animal's responses (lever presses) to changes in hypothalamic or skin temperature (see text for explanation). (Corbit, 1973)

Environmental system. The stimuli, hypothalamic and skin temperatures, were controlled by the system shown in Figure 8-3. Hypothalamic temperature was controlled by controlling the temperature of the water that perfused (at 65 ml per minute) the thermode, and skin temperature was controlled by controlling the temperature of the air that flowed (at 1.0 m per second) through the test chamber and over the animal's body surface. Water and air temperatures were controlled by liquid-to-liquid and liquid-to-air heat exchangers served by constant-temperature circulators. Solenoid-operated valves switched the water and air streams to determine which of two water temperatures perfused the thermode, and which of two air temperatures flowed over the animal's body. The animals could be given control of either hypothalamic or skin temperature by arranging for their responses to activate the valve in the water or air stream. Activation

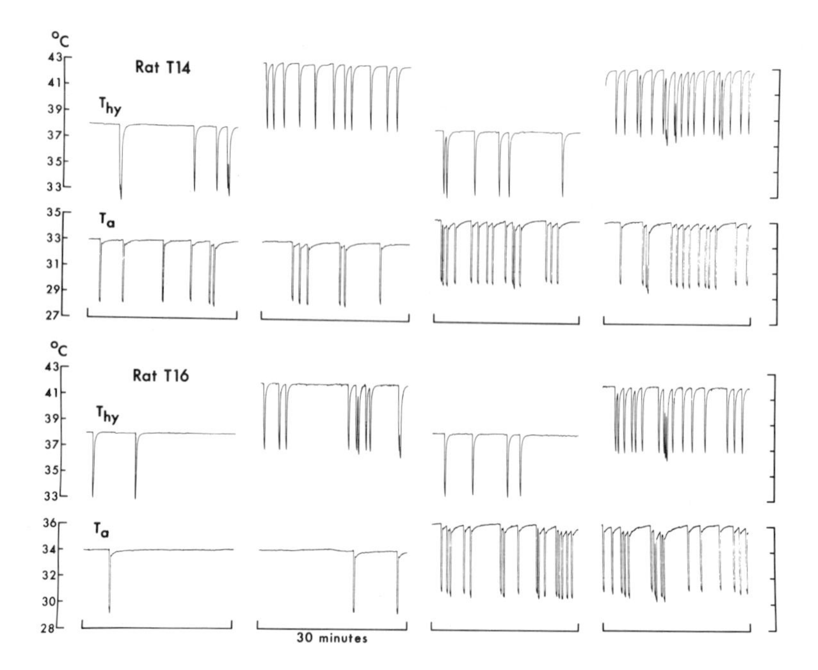

Figure 8-4. Records of hypothalamic (T_{hy}) and air (T_a) temperatures obtained during 30-minute preference tests with two rats. The four test conditions (left to right) involved different combinations of hypothalamic and air temperatures: neutral-neutral, warm-neutral, neutral-warm, and warm-warm. (Corbit and Ernits, 1973)

of one of the valves produced a change in the temperature, but not in the velocity, of the water or air stream, which in turn produced a change in hypothalamic or skin temperature.

Appropriate programming of the temperature-control system defined the relationship between a transduced aspect of the behavioral response and the change in thermal stimulation that it produced. In the present experiments, a very simple input-output relationship was defined for the environmental system—i.e., a sufficient force exerted by the animal on a lever mounted on the wall of the test chamber was detected by a microswitch and converted by the temperature-control system to a brief (15-sec) standard pulse temperature reduction.

Interaction of Central and Peripheral Temperatures

The first series of experiments was undertaken to investigate the interaction of hypothalamic and skin temperatures in determining the intensity of the drive for behavior directed toward reducing air (and therefore skin) temperature. Initially, air temperature was set at a very warm value (38° C), and the rats quickly learned that they could obtain reductions in air temperature by pressing the lever. When the animal pressed the lever, air temperature fell abruptly to a neutral value (30° C), but after 15 sec it returned equally abruptly to the initial warm value where it remained until the animal responded again, whereupon air temperature again fell to neutrality for 15 sec and so on (the waveform of the response-produced changes in air temperature can be seen in Figure 8-4. Under these conditions, the measure of the drive for thermoregulatory behavior was the rate at which the animal responded for the temporary reductions in air temperature. After the animals had sufficient experience with the task, their rates of responding stabilized, and the experiments were begun.

The interaction of hypothalamic and skin temperatures was studied by determining the independent and combined effects of the two stimuli on

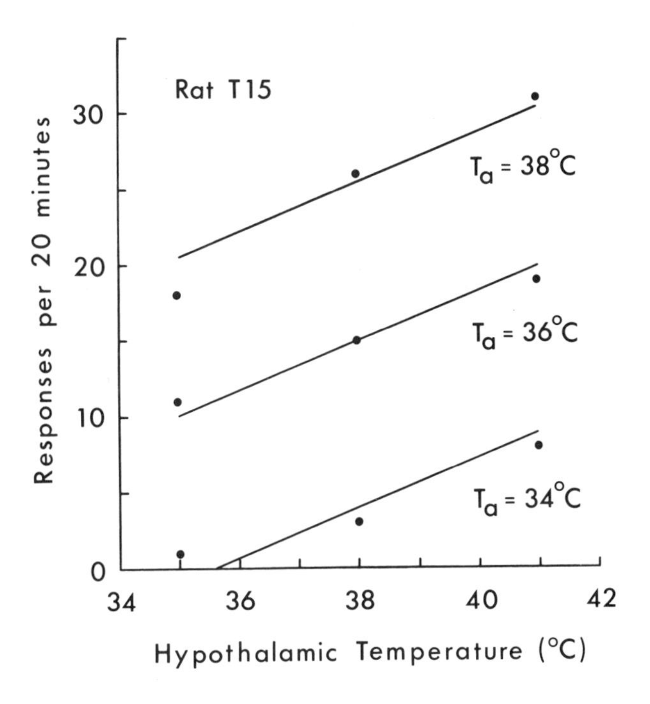

Figure 8-5. Additive interaction of hypothalamic and air temperatures in the control of rate of responding for reductions in air temperature. (Corbit, 1973)

rate of responding. As air temperature was increased above neutrality, rate of responding for reductions in air temperature increased linearly. In addition, rate of responding to reduce a given value of air temperature increased when hypothalamic temperature was raised above its normal value, and decreased when hypothalamic temperature was lowered. The effect of hypothalamic temperature, like that of air (skin) temperature, was linear.

The results obtained from one animal are shown in Figure 8-5, where each data point gives the number of times that the rat called for air-temperature reductions during a 20-minute test period at the indicated values of hypothalamic and air temperatures (skin temperature was proportional to air temperature over the range of warm air temperatures used). Rate of responding increased linearly with increases in hypothalamic temperature and with increases in air (skin) temperature. The results of this and other similar experiments are well summarized by the equation:

$$R = a\,[\,(T_{hy} - T_{hy_0}) + b\,(T_s - T_{s_0})\,] \tag{1}$$

according to which the rate of responding, R (in responses per minute), is proportional to the sum of the weighted deviations of hypothalamic, T_{hy}, and skin, T_s, temperatures from their respective neutral values, T_{hy_0} and T_{s_0} (all temperatures in °C), and where a is the proportionality constant (in responses per minute per °C), and b is a dimensionless weighting constant.

Estimates of the values of the constants in Equation 1 were obtained in the following way. The normal value of hypothalamic temperature for each animal was taken as the estimate of T_{hy_0} (range: 38 to 39° C). To obtain an estimate of T_{s_0}, T_{hy} was held constant at T_{hy_0}, and air temperature was raised and lowered until the threshold for the initiation of responding for reductions in air temperature was found. The threshold value of T_s, given that $T_{hy} = T_{hy_0}$, was taken as the estimate of T_{s_0} (range: 35 to 36° C). The ratio of the effect of a 1° C increase in T_s to the effect of a 1° C increase in T_{hy} provided the estimate of the weighting constant, b, which expresses the relative importance of skin temperature in determining the rate of responding (range: 2 to 3). The effect of a 1° C increase in T_{hy} gave the estimate of the proportionality constant, a (range: 0.05 to 0.25 responses per minute per ° C). Typical values for the constants are given in Equation 2:

$$R = 0.2\,[\,(T_{hy} - 38.5) + 3\,(T_s - 35.5)\,] \tag{2}$$

A Model of the Neural System

Electrophysiological studies of first-order and higher-order temperature-sensitive neurons have defined some of the basic characteristics of the hypothalamic (Hardy, 1969; Hellon, 1970) and cutaneous (Hensel, 1970; Iggo, 1970; Landgren, 1970; Zotterman, 1959) thermal sensory mechanisms. The hypothalamic and cutaneous temperature-sensitive neurons fall into two classes, warm-sensitive and cold-sensitive. The direction (above or below neutrality) of deviations of hypothalamic and skin temperatures from their neutral values is encoded by differential stimulation of the sets of warm-sensitive and cold-sensitive receptors. The extent (how far above or below neutrality) of the temperature deviations is encoded by the frequency of firing by the warm-sensitive and cold-sensitive neurons, whose frequency of firing increases with increasing temperature deviations above and below neutrality, respectively.

Thus, the results of electrophysiological studies of the individual elements of the internal and external sensory mechanisms indicate in a general way how the direction and extent of temperature deviations are encoded as neural signals. Guided by these findings, we make the simplifying assumption that the hypothalamic and cutaneous sensory mechanisms convert thermal stimuli to sensory signals, S_{hy} and S_s, proportional to the differences between the actual receptor temperatures, T_{hy} and T_s, and the neutral or reference temperatures, T_{hy_0} and T_{s_0}, characteristic of the sensory mechanisms. Thus,

$$S_{hy} = (T_{hy} - T_{hy_0}) \tag{3}$$

$$S_s = b\,(T_s - T_{s_0}) \tag{4}$$

where the constant, *b*, gives the effects of a 1° C change in T_s relative to the effects of a 1° C change in T_{hy}; in the rat under the conditions described here, T_s was given about three times as much weight as T_{hy} (Equation 2). Warm ($T > T_0$) and cold ($T < T_0$) stimuli produce proportional sensory signals of positive and negative sign, respectively.

These assumptions must be qualified. First, the assumption of linearity is a reasonable first approximation, but it holds only over a limited range of temperatures, and that range should be specified. Second, Equation 4 ignores the fact that cutaneous receptors are sensitive to the rate of temperature change, as well as to absolute temperature, and therefore holds only for steady-state conditions.

As mentioned above, knowledge of the input-output characteristics of the sensory mechanisms makes it possible to reduce the problem from the single component level of the neural system as a whole as a "black

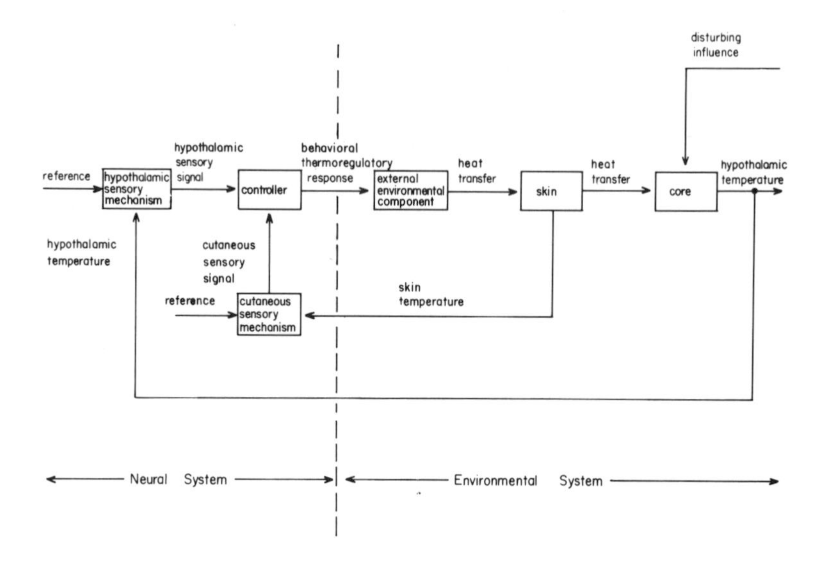

box" to the three-component level of internal and external sensory mechanisms plus a controller. The block diagram model of the generalized behavioral control system (Model 1) can be applied to the specific problem of behavioral temperature regulation, (Model 2). Once the problem is reduced to this three-component level, and the characteristics of the sensory mechanisms are known, the focus of the analysis is on the information-processing characteristics of the controller.

Compare this model, the control system for behavioral temperature regulation, with the generalized system, Model 1. The internal and external stimuli of Model 1 are now hypothalamic and skin temperatures.

The results of the behavioral experiments described above (Figure 8-5) indicate that the neural system for thermoregulatory behavior converts thermal stimuli to behavioral thermoregulatory responses according to the rule given by Equation 1. That is, rate of responding is proportional to the sum of the weighted deviations of hypothalamic and skin temperatures from their respective neutral values. Given knowledge of the information-processing characteristics of the sensory mechanisms, this result can be interpreted in terms of its implications for the information-processing characteristics of the controller. If we assume that the relationships between hypothalamic and skin temperature deviations and the intensities of the hypothalamic and cutaneous sensory signals are linear (Equations 3 and 4), then the behavioral result (Equation 1) implies that the controller is also a linear component, and that it combines the hypothalamic and cutaneous sensory signals by addition to determine the rate of behavioral responding.

Two major types of information-processing operation performed by the controller are discernible. The first type of operation is the integration (additive combination) of sensory information, which determines the intensity with which the behavioral response will be activated. The second type of operation involves the determination of which response (of the many possible responses that could be activated) will be activated. Model 3 shows a block diagram model of the neural system (only) controlling thermoregulatory behavior in which the controller shown in Model 3 has been reduced to two subcomponents. The two distinct functions of the controller, performance of summing and switching operations, are referred to two distinct structural components, an integrator and a response selector.

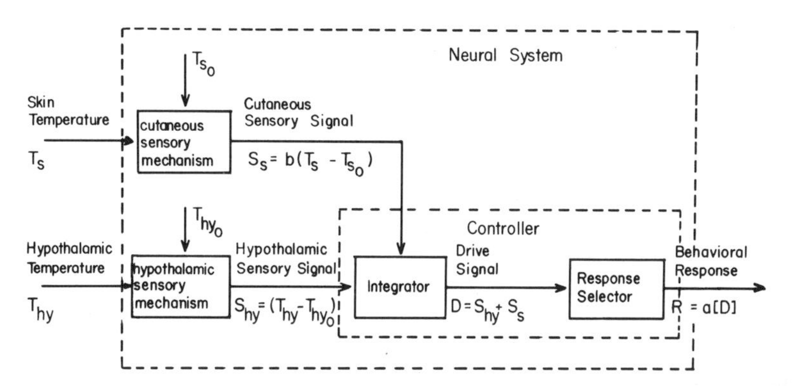

The fact that both hypothalamic and skin temperatures influence a given behavioral thermoregulatory response, implies that the hypothalamic and cutaneous sensory signals converge on an integrative component prior to the final common path for the behavioral response. The integrator in Model 3 is assumed to combine the sensory signals by addition, and to generate as its output a drive signal, *D*, proportional to their sum. Thus,

$$D = S_{hy} + S_s \tag{5}$$

The integrator respects the sign of the sensory signals and adds them algebraically, yielding a drive signal which may be of positive or of negative sign.

The response selector converts the drive signal to behavioral responses. It uses the sign of the drive signal to determine which response will be activated, and it uses the intensity of the drive signal to determine how vigorously it will be activated. Rate of responding, R, is assumed to be proportional to the drive signal. Thus,

$$R = a [D] \quad (6)$$

The sign of the drive signal determines whether a heat-loss or a heat-gain response will be activated. If the drive signal has a positive value, then responses which bring about a reduction in environmental temperature (and therefore a reduction in skin temperature and an increased rate of heat loss) are activated, but if the drive signal has a negative value, responses which bring about an increase in environmental temperature (and an increase in skin temperature and a reduced rate of heat loss) are activated.

To summarize, according to the point of view presented graphically by block diagram Model 3, the neural system consists of hypothalamic and cutaneous sensory mechanisms that convert temperature deviations to sensory signals, an integrator that sums the sensory signals and generates a drive signal, and a response selector that converts the drive signal to behavioral responses. The rules (Equations 3, 4, 5, 6) assumed to govern the transfer of information through the model's sensory, integrative, and response selector components, taken together, yield the rule (Equation 7) found to govern the transfer of information through the actual neural system of the animal.

$$R = a [D] = a [S_{hy} + S_s] = a [(T_{hy} - T_{hy_0}) + b (T_s - T_{s_0})] \quad (7)$$

A review of the behavioral literature shows the results published prior to March 1969 to be in general agreement with Model 3 (Corbit, 1970), and more recent studies have provided quantitative support for the linearity and additivity assumptions (Adair et al., 1970; Corbit, 1969, 1973).

The value of models of this type is that they provide a definite hypothesis with which to organize existing knowledge, and a clearly defined idiom for discussing it. Furthermore, they expose the gaps in existing knowledge and suggest further behavioral and electrophysiological experiments. The behavioral experiments to be described below were undertaken to answer questions raised by Model 3.

Direct Behavioral Control of Hypothalamic Temperature

Normally, animals have direct and immediate behavioral control of external temperature, but the behavioral control of internal temperature is indirect and subject to substantial delays. Nonetheless, despite the fact

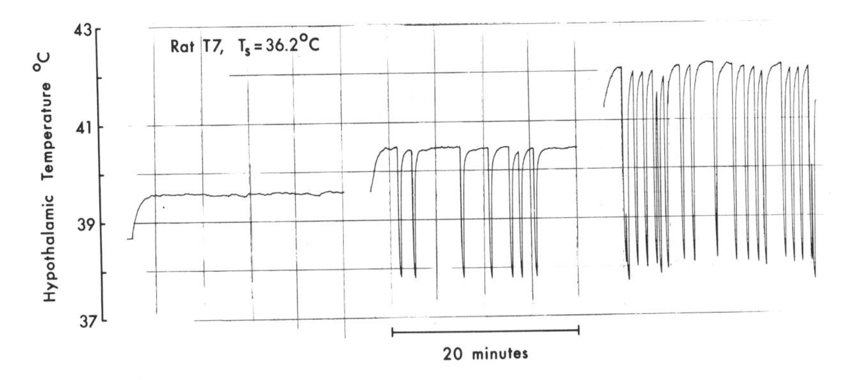

Figure 8-6. Strip-chart records of hypothalamic temperature from three 20-min test periods. Rate of responding for reductions in hypothalamic temperature increased as the steady-state value of hypothalamic temperature was increased. Skin temperature was held constant at 36.2° C during all tests. (Corbit, 1969)

that behavioral control of internal temperature is normally indirect and delayed, the system can be viewed as one whose main function is to compensate for disturbances and maintain internal temperature at its normal value. If the system functions to regulate internal temperature, then when, for example, the hypothalamus is warmed, the animal should not only work to reduce external environmental temperature (Figure 8-5) and thereby achieve an eventual reduction of hypothalamic temperature, but also work directly and immediately to reduce hypothalamic temperature, if provided with a means of doing so.

To test this implication, the environmental system (Figure 8-3) was reprogrammed so that the animal's responses caused 15-sec reductions in hypothalamic temperature, rather than 15-sec reductions in air temperature as before. Then rats with hypothalamic thermodes, which had previously been trained to work for reductions in air temperature in the experiments discussed above, were provided with the opportunity to manipulate hypothalamic temperature directly. After a brief initial period of confusion, the animals exhibited stable rates of responding for reductions in hypothalamic temperature.

Thus, animals will, if the environment permits, work to restore hypothalamic temperature to its normal value. The basic phenomenon is illustrated by Figure 8-6, which presents records of hypothalamic temperature obtained during three 20-minute tests in which hypothalamic temperature was set at progressively higher values, but skin temperature was held constant at 36.2° C. As hypothalamic temperature was increased, the vigor of the animal's efforts to reduce it increased proportionally. Taking rate of responding as a measure of the drive for hypothalamic temperature change, these results support the conclusion that the drive for hypothalamic temperature change, like the drive for skin temperature change (Figure 8-5), is proportional to the deviation of hypothalamic temperature from neutrality.

The drive for hypothalamic temperature change depends on the peripheral temperature, as well as on the hypothalamic temperature itself. The role of peripheral temperature is illustrated by the results of the experiment shown in Figure 8-7. Initially, when air temperature was at a neutral value, there was little or no responding to reduce hypothalamic temperature, even though the hypothalamus was very warm (43.0° C). As air temperature (and therefore skin temperature) was gradually increased from

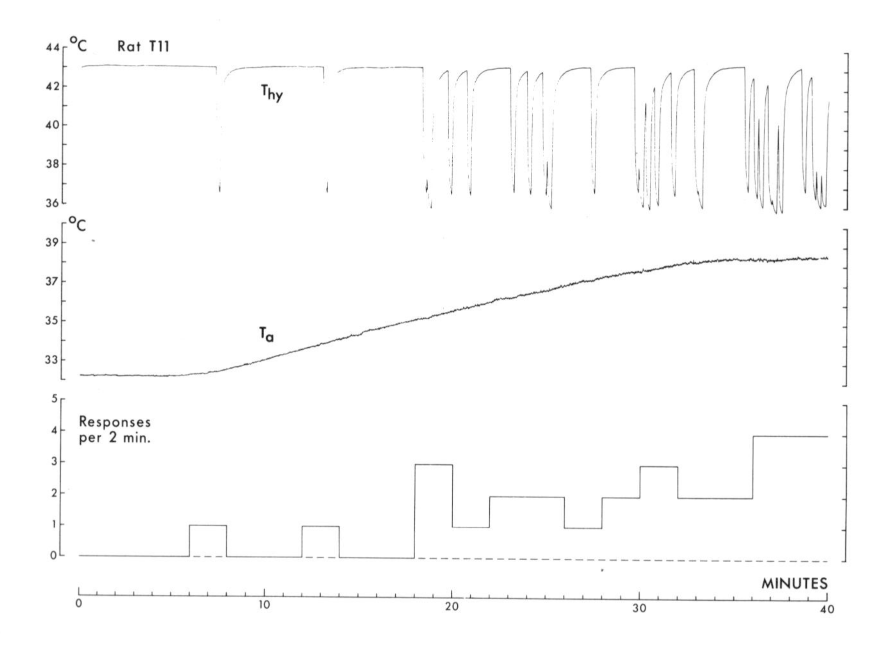

Figure 8-7. Effect of gradual increase in air temperature (T_a) on rate of responding for transient reductions in hypothalamic temperature (T_{hy}). The three records show changes in three aspects of the system: the input stimuli (T_{hy} and T_a), the output of the neural system (rate of responding), and the response-produced feedback (changes in T_{hy}). (Corbit, 1973)

32.2 to 38.5° C, at a rate of about 0.25° C per minute, the rate at which the animal responded for reductions in hypothalamic temperature increased proportionally from 0 to 2 responses per minute. This experiment shows the role of skin temperature, and the system's ability to follow a slowly varying input with a proportional output.

The phenomenon of behavioral control of hypothalamic temperature was examined in greater detail in experiments of the type shown in Figure 8-8, in which continuous records of hypothalamic, rectal, skin, and air temperatures were obtained during the behavioral tests. For the first 20 minutes of the experiment shown in Figure 8-8, hypothalamic and air temperatures were held constant at neutral values of 38.7 and 30.4° C, and the average values of rectal and skin temperatures were about 38.6 and 32.0° C. At 20 minutes hypothalamic temperature was raised to 43.0° C. The hypothalamic warm stimulus elicited cutaneous vasodilation, an automatic heat-loss response, seen as an increase in skin temperature. During the period between 20 and 30 minutes, the increased skin-to-air temperature gradient resulted in an increase in the rate at which heat was lost from the skin surface to the external environment, and a loss of body heat content, seen as a decrease in rectal temperature from

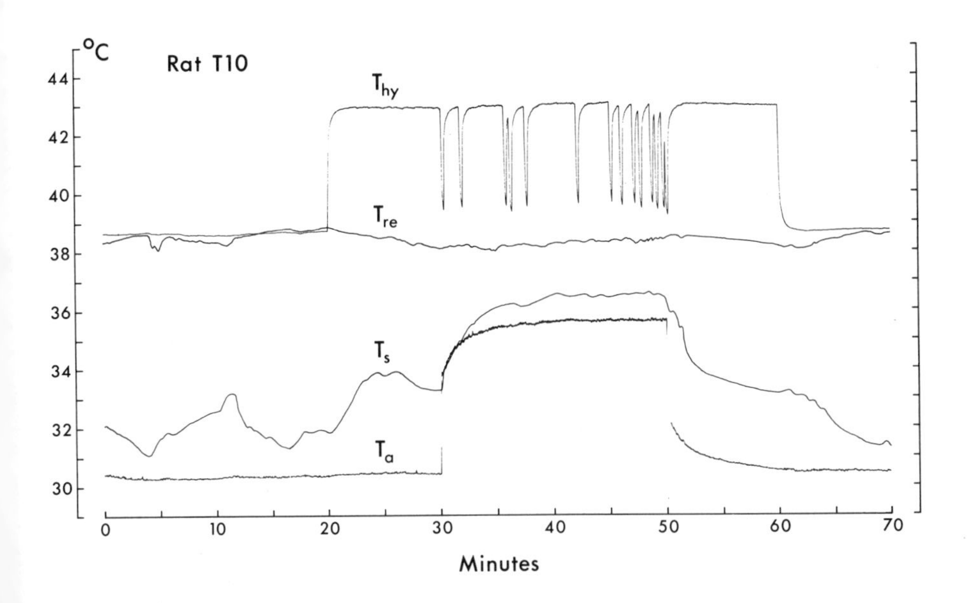

Figure 8-8. Records of hypothalamic (T_{hy}), rectal (T_{re}), skin (T_s), and air (T_a) temperatures (see text for explanation). (Corbit, 1973)

38.8 to 38.1° C. Thus, when air temperature was at a neutral value, there was a clear automatic vasomotor response but no behavioral response to the increase in hypothalamic temperature.

At 30 minutes, air temperature was increased from 30.4 to 35.6° C, causing a further increase of skin temperature from 33.3 to 36.5° C, whereupon the combined effects of the hypothalamic and cutaneous warm stimuli were sufficient to activate behavioral responding, which continued until air and skin temperatures were reduced at 50 min. During the period of elevated air temperature, the skin-to-air temperature gradient and thus the rate of heat loss were reduced, and rectal temperature, which had been falling, began to increase from 38.1° C at 30 minutes to 38.5° C at 50 minutes. With the reduction in air temperature at 50 minutes, skin temperature also fell, but since the skin was still vasodilated under the influence of the warm hypothalamus, the skin-to-air temperature gradient increased; rectal temperature began to decline. At 60 minutes, the system had returned to the same physical state seen at 30 minutes. At this point, hypothalamic temperature was returned to the initial neutral value of 38.7° C, resulting in cutaneous vasoconstriction and a rise in rectal temperature. By 70 minutes, the system had recovered the initial conditions seen during the first 20 minutes, before hypothalamic temperature was raised.

The experiments shown in Model 3 and Figures 8-7 and 8-8 help to characterize the phenomenon of behavioral control of hypothalamic temperature, and specifically show that the drive for hypothalamic temperature change depends on the values of both hypothalamic and skin temperatures. The next step was to show precisely how the two stimuli interact to determine the rate of responding for hypothalamic temperature reductions. This was done by systematically varying hypothalamic and skin temperatures, and recording the resulting changes in response rate, as was done in the analysis of the interaction of the two stimuli in the control of behavioral responding for skin temperature reductions described above (Figure 8-5).

On a given day, air temperature was set at a value between 33 and 39° C. After skin and rectal temperatures had stabilized, hypothalamic temperature was set at different values, and rate of responding was recorded. On subsequent days this procedure was repeated at different air temperatures, yielding relationships between rate of responding and hypothalamic temperature, for each value of skin temperature.

Figure 8-9 presents the results of a detailed study of one animal in which rate of responding, as a function of hypothalamic temperature, was determined at each of four values of skin temperature. Each data point plotted in Figure 8-9 gives the number of times that the animal responded to reduce hypothalamic temperature during a 20-min test period. When skin temperature was 36.2° C, rate of responding increased linearly as hypothalamic temperature was increased above 38.5° C (records of hypothalamic temperature during three tests at T_s, 36.2° C are shown in Figure 8-6). Rate of responding was a linear function of hypothalamic temperature, with the same slope at all values of skin temperature.

The effect of skin temperature was also linear; increasing or decreasing skin temperature above or below 36.2° C resulted in a proportional increase or decrease in the rate at which the animal responded to reduce a given value of hypothalamic temperature. When skin temperature was 38.3° C, the animal called for reductions with any value of hypothalamic temperature greater than 32.3° C. That is, when skin temperature was very warm, the animal vigorously worked to reduce values of hypothalamic temperature which were already far on the cool side of neutrality. On the other hand, at the lowest value of skin temperature studied, 34.5° C, the animal did not begin to call for hypothalamic temperature reductions until hypothalamic temperature had exceeded 42.3° C.

It is clear from the results shown in Figure 8-9, and those obtained from six other rats studied similarly (Corbit, 1969, 1973), that re-

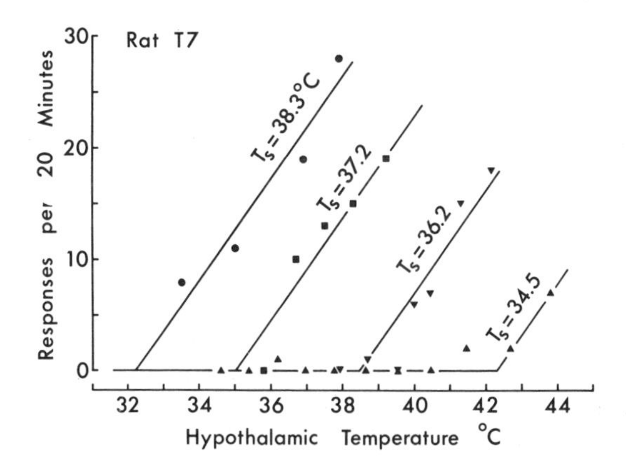

Figure 8-9. Additive interaction of hypothalamic and skin temperatures in the control of rate of responding for transient reductions in hypothalamic temperature. Each data point is based on a 20-min test at the indicated combination of hypothalamic and skin temperatures. The different symbols identify tests conducted at different values of skin temperature. (Corbit, 1969)

sponding for transient reductions in hypothalamic temperature is governed by the same type of additive law (Equation 1) found to characterize the interaction of hypothalamic and skin temperatures in the control of responding for reductions in skin temperature (Figure 8-5). Rate of responding for reductions in either temperature is proportional to the sum of the weighted deviations of hypothalamic and skin temperatures from their respective neutral values. Furthermore, not only are the rules for combining stimuli in the control of the two types of response of the same form (Equation 1), they are also quantitatively similar. The estimates of the constants in Equation 1, based on data from experiments of the type shown in Figure 8-9, yielded values within the same ranges of values found using data from experiments of the type shown in Figure 8-5 (i.e., $0.05 < a < 0.25$ responses per minute per $^\circ$ C; $2.0 < b < 3.0$; $38 < T_{hy_0} < 39^\circ$ C; and $35 < T_{s_0} < 36^\circ$ C). The typical values of the constants, given in Equation 2, apply equally well to the control of responding for reductions in hypothalamic temperature as to the control of responding for reductions in skin temperature.

Drive-Signal Minimization

According to the view of the neural system presented graphically as a block diagram (Model 3), hypothalamic and cutaneous sensory mechanisms convert temperature deviations to proportional sensory signals (Equations 3 and 4), the sensory signals converge on an integrator which combines them to generate a drive signal related to their sum (Equation 5), and the response selector converts the drive signal to a proportional rate of behavioral responding (Equation 6).

This model was suggested as a way of thinking about the mechanism responsible for the additive interaction of hypothalamic and skin temperatures in the control of responding for reductions in skin temperature (Figure 8-5; Equations 1 and 2). A review of the model should take into account the results of the studies of responding for reductions in hypothalamic temperature, as well as in skin temperature. The experiments described above showed first of all that, if provided with a means of doing so, animals will directly manipulate hypothalamic temperature, and further that the rate of responding for hypothalamic temperature reductions depends on the value of hypothalamic temperature (Figure 8-6), and skin temperature (Figures 8-7 and 8-8), and that when both stimuli are varied their effects are additive (Figure 8-9; Equations 1 and 2). These new results are completely consistent with Model 3, and they also enable further specification of the nature of the information-processing operations performed by the integrator and response selector components of the model.

The additive interaction of hypothalamic and skin temperatures, in the control of responding for hypothalamic temperature reductions (Figure 8-9), provides further support for the idea that the operation performed by the integrator is summation—i.e., that the integrator combines the sensory signals and generates a drive signal related to their sum (Equation 5). Furthermore, the phenomenon of behavioral control of hypothalamic temperature is not a subtle or weak effect. Actually, for a given set of stimulus conditions, responding for 15-sec reductions in hypothalamic temperature and responding for 15-sec reductions in skin temperature are about equally vigorous—i.e., given by the same equation (Equation 2). In view of this quantitative similarity, the simplest assumption is that there is only one integrator, and that both types of response are driven by a common drive signal.

Now, once we have some idea of how the integrator combines sensory information to generate the drive signal, the next question is, given a drive signal, how is it converted to behavioral responses by the response selector? Specifically, what are the general principles that determine which particular response will be activated, and, given that the response has been selected for activation, how vigorously it will be activated?

The latter quantitative question has been dealt with by assuming that the vigor of the behavior is related to the intensity of the drive signal, and specifically that, under the conditions of the experiments described here, rate of responding is proportional to the intensity of the drive signal (Equation 6). This assumption is based on the idea that an intense drive signal results in the activation of the response more rapidly than does a weak drive signal—i.e., that one characteristic of the response selector is an intensity-duration relationship, such that the latency of the response (time from drive signal onset to the occurrence of the response which reduces the drive) is inversely related to drive-signal intensity.

The qualitative question of which particular response will be selected for activation can be answered to this extent. We observe that animals always behave in such a way as to reduce the drive signal, and that they will manipulate either hypothalamic or skin temperature to bring about a reduction in the value of the drive signal, and we infer from this that one major principle governing the operation of the response selector is drive-signal minimization—i.e., the response selector apparently always selects, for activation by the drive signal, those responses which have as their consequences reductions in the absolute value of the drive signal.

The evidence for drive-signal minimization is as follows. When hypothalamic temperature is displaced from neutrality, the system responds to reduce the resulting drive signal either by returning hypothalamic temperature to neutrality (Figure 8-6) or by displacing skin temperature from neutrality in the opposite direction (Adair et al., 1970; Carlisle, 1966; Corbit, 1970; Murgatroyd and Hardy, 1970; Satinoff, 1964); when skin temperature is displaced from neutrality, the system responds to reduce the resulting drive signal either by returning skin temperature to neutrality (Corbit, 1970; Matthews, 1971; Weiss and Laties, 1961) or by displacing hypothalamic temperature from neutrality in the opposite direction (Figure 8-9).

The fact that when, for example, the skin is warmed, animals work to return skin temperature to neutrality shows that the system tends to minimize the value of the cutaneous sensory signal. However, it minimizes the cutaneous sensory signal only when to do so is consistent with drive-signal minimization. When cutaneous sensory signal minimization and drive-signal minimization are in conflict, drive-signal minimization prevails. For example, when the hypothalamus is warmed and it is not possible to return hypothalamic temperature to neutrality, animals work to displace skin temperature from neutrality to a cool value, thereby increasing rather than decreasing the cutaneous sensory signal, but reducing the drive signal. Thus, the system gives higher priority to drive-signal minimization than to cutaneous sensory signal minimization.

Similarly, the fact that, when the hypothalamus is warmed, animals work to return hypothalamic temperature to neutrality (Figure 8-6) is evidence for a tendency to minimize the hypothalamic sensory signal. But even the hypothalamic sensory signal is minimized only when its minimization is consistent with drive-signal minimization. Again, when sensory signal minimization and drive-signal minimization are in conflict, drive-signal minimization prevails. When the skin is very warm and it is not possible to return skin temperature to neutrality, animals work to change hypothalamic temperature from a neutral value to a cool value (Figure 8-9), thereby increasing rather than decreasing the hypothalamic sensory signal but, as always, reducing the drive signal. Thus, the system gives higher priority to drive-signal minimization than to hypothalamic sensory signal minimization.

To summarize, it is clear that when one temperature is displaced from neutrality, the system responds to return that temperature to neutrality, and when this direct approach is not possible, the system responds to move the other temperature away from neutrality in the opposite direction. Apparently the system gives highest priority to drive-signal minimization, and the maintenance of hypothalamic and skin temperatures at their respective neutral values (i.e., sensory signal minimization) is of secondary importance.

Sensory Signal Minimization

The experiments on behavioral control of skin temperature and on behavioral control of hypothalamic temperature indicate that the neural system for thermoregulatory behavior operates to minimize the value of the drive signal. A drive signal, established by displacing one or both temperatures from neutrality, results in the activation of responses which bear a negative feedback relationship to the drive signal. Animals will adjust either skin temperature or hypothalamic temperature in such a way as to reduce the absolute value of the drive signal, and the system apparently becomes quiescent when it has achieved pairs of T_{hy} and T_s, such that $(T_{hy} - T_{hy_0}) + b(T_s - T_{s_0}) = D = 0$.

A drive signal established by displacing one of the temperatures can be reduced, either by returning the displaced temperature to neutrality or by displacing the other temperature away from neutrality in the opposite direction. We have seen, for example, that warming the skin causes a proportional increase in behavioral responding to return skin temperature to neutrality (Figure 8-5), and in responding to reduce hypothalamic temperature to cool values below neutrality (Figure 8-9). In both cases the drive is reduced to zero. But, in the former case both temperatures reach neutrality ($T_{hy} = T_{hy_0}$ and $T_s = T_{s_0}$), whereas in the latter case neither temperature reaches neutrality ($T_{hy} < T_{hy_0}$ and $T_s > T_{s_0}$). These endpoints are equivalent from the point of view of drive-signal minimization, but they are not equivalent either at the level of sensory signals or from the point of view of temperature regulation. There is a definite thermoregulatory advantage to minimizing the sensory signals by behaving in such a way that both temperatures end up at neutrality. Therefore, experiments were undertaken (Corbit and Ernits, 1974) to find whether the system shows a preference for minimizing the drive signal by reducing both sensory signals to zero, as opposed to minimizing the drive signal by establishing sensory signals of equal value and opposite sign.

Rats were trained to respond one way (press a lever) to produce 15-sec reductions in hypothalamic temperature, and to respond another way (pull a chain) to produce 15-sec reductions in skin temperature. During the preference tests both the chain and the lever were made available, and the animals were free to reduce either hypothalamic or skin temperature.

Prior to the preference tests, estimates of the neutral values of hypothalamic (T_{hy_0}) and skin (T_{s_0}) temperatures were obtained. Hypothalamic temperature was normally about 38° C in the animals used in these experiments, and 38° C was taken as the value of T_{hy_0}. The value of T_{s_0} was found by raising and lowering air temperature, while

Table 8-1. Preferential Responses to Air (T_a) and Hypothalamic (T_{hy}) Temperature Changes

Rat	Stimuli (°C)		Responses per 30 minutes	
	T_a	T_{hy}	$\downarrow T_a$	$\downarrow T_{hy}$
T-14	33	38	6	7
	33	43	6	11
	35	38	15	6
	35	43	14	19
T-16	34	38	1	2
	34	42	2	10
	36	38	16	4
	36	42	15	17

T_{hy} was held constant at T_{hy_0} = 38° C until a rough estimate of the threshold for responding for reductions in air temperature was obtained. The threshold values of air temperature varied between 33 and 34° C, and the value of T_{s_0} was about 35° C.

Drives for temperature decrease were established by 4 or 5° C increases in hypothalamic temperature, or by 2° C increases in skin temperature, or by increases in both temperatures. Pressing the lever produced 5° C reductions in hypothalamic temperature; pulling the chain produced 5° C reductions in air temperature, which reduced skin temperature by about 2° C.

The results of preference tests with two rats are presented in Figure 8-4 and Table 8-1. The figure shows records of hypothalamic and air temperatures obtained during four 30-minute tests of each animal, and the table gives the number of times that the rats pressed the lever and pulled the chain during each 30-minute test period. When both temperatures were at neutrality, there was little of either type of responding. When air temperature was increased and hypothalamic temperature was at neutrality, rate of responding for reductions in air temperature increased, but rate of responding for reductions in hypothalamic temperature did not change. When hypothalamic temperature was increased and air temperature was at neutrality, rate of responding for reductions in hypothalamic temperature increased, but rate of responding for reductions in air temperature did not change. When both temperatures were increased, both response rates increased.

The results of the preference experiments are consistent with the idea that the drive for thermoregulatory behavior is generated by additive combination of the hypothalamic and cutaneous sensory signals, and that the system operates to reduce the drive, but they are not consistent with the idea that all means of drive-signal reduction are equivalent. When free to operate on both temperatures, the animal prefers to reduce the drive signal by returning the displaced temperature to neutrality, thereby reducing sensory as well as drive signals. The alternative approach to drive-signal reduction (nulling out the effects of warming one receptor site by cooling the other receptor site) is not preferred, although, as previous experiments have shown, this approach is used when no other means of reducing the drive is possible.

Thus, the system is capable of more than drive-signal minimization. It is capable of distinguishing between changes in hypothalamic temperature and changes in skin temperature, and when the external environment provides the necessary options for control of the stimuli, the neural system minimizes the sensory signals as well as the drive signal.

Discussion and Conclusions

Summary of Findings

The main conclusions to be drawn from the series of studies of thermoregulatory behavior in the rat are: (1) hypothalamic and skin temperatures are stimuli for thermoregulatory behavior; (2) each stimulus has a linear effect—rate of responding varies linearly with deviations of either stimulus from its neutral value; (3) when both stimuli are varied, their effects are independent and additive (Figures 8-5 and 8-9)—rate of responding is proportional to the sum of the weighted deviations of hypothalamic and skin temperatures from their respective neutral values (Equations 1 and 2); (4) skin temperature is given about three times as much weight as hypothalamic temperature in determining rate of responding—the effect on rate of responding of a 1° C increase in skin temperature is about three times as great as the effect of a 1° C increase in hypothalamic temperature; (5) the animal will work to change either hypothalamic or skin temperature and, under a given set of stimulus conditions, responds at about the same rate for reductions in hypothalamic temperature as for reductions in skin temperature—the same equation (Equation 2) describes rate of responding for both types of temperature change; (6) drive-signal minimization is given higher priority than sensory signal minimization—the animal always responds so as to reduce the absolute value of the drive signal, even when to reduce the drive signal it must move the temperature under its control away from rather than toward neutrality; (7) if possible, the animal, will minimize the sensory signals as well as the drive signal—when the animal is free to operate on both temperatures, it preferentially operates on the temperature which has been displaced and returns it to neutrality (Figure 8-4, Table 8-1).

Updating the Model

Model 3 of the neural system can account for the results of the studies of the interaction of hypothalamic and skin temperatures in the control of responding for changes in skin temperature (Figure 8-5), and in the control of responding for changes in hypothalamic temperature (Figure 8-9), but it can not account for the results of the preference experiments (Figure 8-4, Table 8-1).

The response selector in Model 3 receives only one input, the drive signal, and thus the only information available to be taken into account in the process of response selection is that contained in the drive signal. The drive signal contains information regarding the severity (its intensity) and the direction (its sign) of the thermal disturbance, but no information regarding the origin (central or peripheral) of the disturbance.

In a neural system of this type, the response selector could use the intensity of the drive signal to determine the vigor of the behavior, and the sign of the drive signal (positive or negative, depending on whether warm or cold stimulation is predominant) to determine whether to activate a temperature-reducing (heat-loss) or temperature-increasing (heat-gain) response. By activating temperature-reducing responses when the drive signal has a positive sign, and temperature-increasing responses when the drive signal has a negative sign, the system would achieve drive-signal minimization. But since the drive signal is the only input to the response selector, and depends on both hypothalamic and skin temperatures, it provides no basis for deciding which temperature has been displaced, and therefore no basis for preferential responding for changes in one temperature as opposed to the other. The neural system shown in Model 3 is thus capable of drive-signal minimization, but not of sensory signal minimization. It regards all pairs of T_{hy} and T_s, such that $D = 0$, as equivalent endpoints, and, in the preference situation it is just as likely to be found with warm brain and cool skin, as with neutral brain and neutral skin temperatures.

The actual neural system of the animal is, however, able to discriminate internal temperature change from external temperature change, and is capable of sensory signal minimization as well as drive-signal minimization. In the preference experiments, where they were free to reduce either hypothalamic or skin temperature, the animals preferentially reduced hypothalamic temperature when hypothalamic temperature had been raised, and preferentially reduced skin temperature when skin temperature had been raised. The preferential reduction of the displaced temperature (i.e., sensory signal minimization) implies that information regarding the origin, as well as the severity and direction, of the thermal disturbance must be present at the level of response selection. More specifically, the response selector must receive an input from at least one of the sensory mechanisms as well as from the integrator.

A new version of the model, modified in an effort to accommodate the results of the preference experiments, is shown below. The updated model (Model 4) is similar to Model 3 in that it involves three stages of linear information processing, performed by sensory, integrative, and response selector components. Again, we assume (1) that the sensory mechanisms convert central and peripheral thermal stimuli to proportional sensory signals that contain information regarding the extent and direction of deviations of hypothalamic and skin temperatures from their respective neutral values (Equations 3 and 4), (2) that the sensory signals converge on an integrator which combines them by addition to generate a drive signal proportional to their sum (Equation 5), and (3) that the drive signal is converted to a proportional rate of behavioral responding by the response selector (Equation 6). The new model differs from the earlier version (Model 3) only in that now the response selector receives an input from the cutaneous sensory mechanism as well as from the integrator, and in that now the response selector has two outputs (R_1 and R_2, i.e., lever press and chain pull).

All of the evidence available at present is consistent with the idea that the drive signal is the sole determinant of the vigor of the behavioral

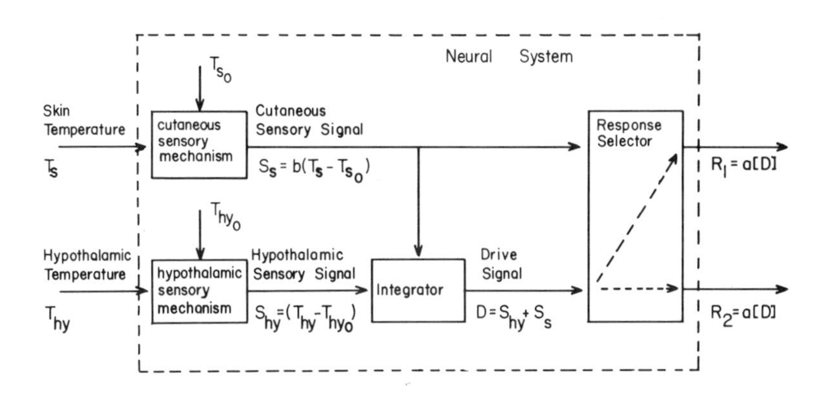

response, and that the sensory signal input to the response selector serves a purely guiding function (i.e., contributes only to the determination of which response will be activated, and not to the determination of the vigor with which it is activated). Assuming that this idea is correct, the response selector can be thought of as behaving like a relay which switches the drive signal to activate one of two responses. The broken arrows inside the response selector in Model 4 are intended to represent this concept of switching the drive signal, controlled by information regarding the origin of the thermal disturbance provided by the cutaneous sensory signal. The "relay" might be operated by the cutaneous sensory signal as follows: if there is a cutaneous sensory signal, then the drive signal is switched to activate the response which results in a change in skin temperature; if there is no cutaneous sensory signal, then the drive signal is switched to activate the response which results in a change in hypothalamic temperature.

A mechanism of this type would account for the animal's ability to discriminate hypothalamic temperature change (change in the drive signal, but no change in the cutaneous sensory signal) from skin temperature change (change in both the drive and cutaneous sensory signals). This system, in contrast to the one shown in Model 3, does not regard all pairs of T_{hy} and T_s, such that $D = 0$, as equivalent endpoints. Instead, it minimizes the sensory signals as well as the drive signal, and tends to end up at $T_{hy} = T_{hy_0}$ and $T_s = T_{s0}$.

The neural system in Model 4 can account for the preferential responding under the warm-neutral and neutral-warm conditions of the preference experiments, but it gives a much poorer account of the pattern of behavior seen under the warm-warm condition. If the operation of the "relay" depended only on skin temperature, then when both hypothalamic and air (skin) temperatures were elevated, the animals should have worked almost exclusively to reduce air temperature. This neural system also predicts that under the warm-warm condition the animal would respond to reduce hypothalamic temperature only during the 15 sec immediately after its response to reduce air temperature. Only during the 15-sec periods when air temperature is at a neutral value would there be a positive drive signal and no cutaneous sensory signal, and only then would the conditions for activating the hypothalamic temperature-reducing response be met. The pattern of responding shown under the warm-warm condition (Figure 8-4) does not support these predictions of this new model: the animals worked vigorously to reduce both temperatures, and they frequently responded to reduce hypothalamic temperature at times when air temperature was elevated. This finding suggests that the operation of the "relay" depends on more than just skin temperature, and that hypothalamic as well as skin temperature information must be present at the level of response selection. The "relay's" switching of the drive signal to activate hypothalamic temperature-reducing responses is determined by the hypothalamic sensory signal, and not simply by the absence of a cutaneous sensory signal. The new model above is too simple.

A final version, Model 5, is shown below. Model 5 is identical to Model 4 except that now the response selector receives an input from the hypothalamic sensory mechanism as well as inputs from the cutaneous sensory mechanism and from the integrator. Like Model 4, Model 5 minimizes sensory signals as well as the drive signal, and tends to end up at $T_{hy} = T_{hy_0}$ and $T_s = T_{s_0}$, and it can account for the preferential responding to reduce the displaced temperature seen under the warm-neutral and neutral-warm conditions; unlike Model 4, it can also account for the pattern of responding seen under the warm-warm condition.

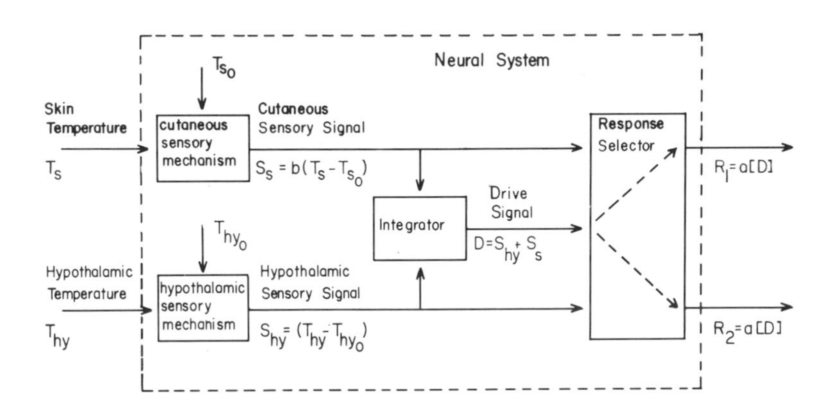

Even Model 5 is certainly much too simple a view of the neural system. First of all, it considers only two inputs, hypothalamic and skin temperatures. But actually there are many factors which influence thermoregulatory behavior, including the temperatures of the caudal brainstem (Lipton, 1971) and of the abdominal viscera (Adair, 1971), the presence of pyrogens (Cabanac, 1969; Cabanac et al., 1970), and other, nonthermal factors—e.g., behavior (Corbit, 1970). The model is also clearly too simple in that it makes no assumptions about some of the major aspects of brain function in relation to the control of voluntary behavior: it considers the sensory, the motivational, and the most elementary type of decision-making mechanisms, but it does not consider the mechanisms of attention, perception, memory, or thinking. Recognition of these ways in which the model is incomplete should not detract from its present value, but rather indicate ways in which it can be made more general in the future.

Despite its simplicity, the model is of value as a way of organizing the results not only of behavioral studies of the sensory and motivational effects of thermal stimulation in animals, but also of psychophysical studies of the sensory and motivational effects of thermal stimulation in man. The intensity of the sensation of warmness or coolness, referred to the skin, increases with increasing deviation of skin temperature from its neutral value (Gagge et al., 1967; Stevens and Stevens, 1963). The intensity of the feeling of thermal discomfort increases with increasing deviation of skin temperature from its neutral value (Gagge et al., 1967; Hardy, 1954; Winslow et al., 1937) and with increasing deviation of internal body temperature from its neutral value (Cabanac, 1969; Chatonnet and Cabanac, 1965). If we assume that the intensity of the sensation of warmness or coolness is related to the intensity (and sign) of the cutaneous sensory signal and that the intensity of the feeling of thermal discomfort is related to the intensity of the drive signal, then the model provides at least a qualitative account of the effects of changes in internal and external temperatures on two dimensions of human experience (i.e., warmness-coolness and comfort-discomfort).

Thermoregulatory Significance

The problem of physiological thermoregulation. What are the thermoregulatory implications of our behavioral findings? First, to establish the general context for these implications, some of the main features of the problem of physiological temperature regulation will be briefly reviewed. Our treatment will be greatly simplified (for thorough reviews of different aspects of this problem see Bligh, 1966; Chatonnet and Cabanac, 1965; Hammel, 1966, 1968; Hardy, 1954, 1961, 1967, 1969; Thauer, 1964, 1970).

In mammals, physiological temperature regulation (i.e., the maintenance of a relatively constant and optimal value of internal body temperature in the face of variations in the heat load on the body) is accomplished by the combined activities of behavioral and automatic thermoregulatory systems. To appreciate the respective contributions to thermal homeostasis made by behavioral and automatic thermoregulatory activities, it is worth first examining what happens in their absence, that is, first considering the body as a passive physical object without active thermoregulatory ability.

We begin by noting that heat is produced continuously in the body as a by-product of metabolism. The metabolic heat produced in the interior of the body is transferred to the body surface and lost to the external environment. If we assume that the body is a passive physical object with constant heat-transfer characteristics and in which heat is produced at a constant rate, then in the steady state, heat is lost to the environment at precisely the rate at which it is produced internally, and internal body temperature stabilizes at a value which depends on the value of the external environmental temperature. We will call the value of internal body temperature reached in the steady state by a passive body, the passive equilibrium value of internal body temperature. This value can be above, below, or identical to the otpimal value of internal body temperature, depending on the value of external environmental temperature.

In a physiologically neutral thermal environment, by definition of physiological neutrality, the passive equilibrium value of internal body temperature reached in the steady state is identical to the optimal value of internal body temperature. Therefore, in a neutral environment, the problem of physiological temperature regulation (i.e., of maintaining internal body temperature at the optimal value) does not arise, because the optimal value is reached and maintained as a purely passive consequence of the physics of the situation, and no active thermoregulation is required. On the other hand, in warm or cool environments the passive equilibrium value of internal body temperature is above or below the optimal value and, in the absence of active thermoregulation, internal body temperature would rise or fall to approach the nonoptimal passive equilibrium value in the steady state.

Now, having defined the concept of physiological neutrality, we can say that the essence of behavioral temperature regulation is the selection of a neutral thermal environment. The behavioral solution to the problem posed by a warm or cool environment is simply to leave that environment and move to a neutral one. By locating themselves in a neutral environment, animals can achieve a rather complete and lasting solution to the problem of physiological temperature regulation.

All vertebrates employ behavioral temperature regulation—i.e., tend to locate themselves in a neutral environment. This behavioral tendency is the only mechanism of active thermoregulation shown by ectotherms (fish, amphibia, and reptiles). These animals lack the ability to alter the rates of heat production and heat loss in response to changing environmental demands, and can only maintain internal body temperature at the optimal value by remaining in a neutral environment. When these animals leave the neutral environment, as they may when driven by influences not directly related to temperature regulation (e.g., to forage for food, to escape predators, etc.), the passive equilibrium value for internal body temperature is shifted above or below the optimal value, and internal body temperature rises or falls to approach the new nonoptimal passive equilibrium value, with a time course determined by the physical properties of the body and the environment.

Behavioral selection of thermal neutrality is also the major thermoregulatory mechanism in endotherms (birds and mammals). However, the endotherms are not quite so dependent on external environmental conditions as are the ectotherms, because they possess, in addition to the behavioral thermoregulatory system that adjusts heat sources and heat sinks in the external environment, an automatic thermoregulatory system that controls the activities of metabolic vasomotor, and evaporative effectors within the body. If the endotherm is unable to locate itself in a neutral environment, temperature regulation is not necessarily compromised, because automatic adjustments of heat production and heat loss can compensate for the heat loads imposed by a limited range of environmental temperatures on the warm and cool sides of neutrality. In a warm environment, the automatic system can maintain internal body temperature near the optimal value by reducing the rate of metabolic heat production (inactivity), increasing the rate of heat loss by increasing the convective transfer of heat to the body surface (peripheral vasodilation), and increasing the rate of evaporative cooling of the body surface (sweating and panting). In a cool environment, the automatic system can maintain internal body temperature near the optimal value by increasing the rate of heat production (shivering and nonshivering thermogenesis), and reducing the rate of heat loss by reducing the flow of warm blood to the body surface (peripheral vasoconstriction).

Thus, by adjusting the rates of heat production and heat loss, the automatic system can successfully oppose a limited range of positive and negative heat loads imposed by warm and cool environments, and can maintain a relatively constant value of internal body temperature more or less indefinitiely. However, in more extreme hot or cold environments, where the heat loads on the body exceed the capacity of the system to lose or produce heat, the automatic system can not maintain thermal balance, and internal body temperature rises above or falls below the optimal value. Therefore, it is absolutely essential that the behavioral system function in such a way that the animal escapes from extreme thermal environments, and remains within the limited range of favorable thermal environments where regulation of internal body temperature by the automatic system is possible. In addition to escaping from environmental extremes, which is obligatory, the behavioral system also tends to seek thermal neutrality, which, while optional from the point of view of survival, nonetheless makes an important contribution to the general physiological economy. To the extent that the animal is successful in locating itself in a neutral environment, thermal balance is achieved passively, minimal levels of automatic thermoregulatory activity are called for, and the overall metabolic cost of maintaining thermal homeostasis is minimized.

Thermoregulatory implications of the model. The results of the behavioral experiments (Figures 8-4, 8-5, and 8-9) indicate that the neural system (final model) operates always to minimize the drive signal and whenever possible to minimize the sensory signals, and tends to end up with $T_{hy} = T_{hy_0}$, T_{s_0}, and $D = 0$. To evaluate how a neural system with these properties would be expected to contribute to the regulation of internal body temperature, first it is necessary to make definite assumptions about how the neutral temperatures for the sensory mechanisms (T_{hy_0} and T_{s_0}) are related to steady-state body temperatures in a neutral environment. The results of electrophysiological studies of the sensory mechanisms (Hardy, 1967, 1969; Hellon, 1970; Hensel, 1970; Iggo, 1970; Zotterman, 1959) support the idea that the hypothalamic and cutaneous sensory mechanisms are "tuned" such that T_{hy_0} is equal to the optimal value of internal body temperature and T_{s_0} is equal to the value of skin temperature reached in the steady-state by a resting subject in a neutral thermal environment.

The assumptions about the "tuning" of the sensory mechanisms, together with the fact that the system tends to end up with $T_{hy} = T_{hy_0}$, $T_s = T_{s_0}$, and $D = 0$, imply that the behavioral system provides a more or less complete solution to the problem of physiological temperature regulation. That is, they imply that an animal, equipped with the neural system shown in Model 5 would tend to locate itself in a neutral thermal environment ($T_s = T_{s_0}$), where internal body temperature would be passively maintained at the optimal value ($T_{hy} = T_{hy_0}$), and therefore where minimal levels of automatic thermoregulatory activity are required for fine-control of internal temperature, and where the cost of achieving thermal homeostasis is at a minimum.

The long-term tendency of the system to maintain itself in a neutral thermal environment is the result of the moment-to-moment behavioral responses to external and internal disturbances. When external environmental temperature deviates from neutrality, the system responds by moving from the nonneutral to a neutral environment. When internal body temperature deviates from the optimal value, the system responds by moving from the neutral to a nonneutral environment such that the return of internal temperature to the optimal value is speeded, and when internal temperature has returned to the optimal value, the system moves back to the neutral environment. For example, a hyperthermic animal would behave in such a way as to reduce external temperature until the effects of cutaneous cold stimulation balanced the effects of hypothalamic warm stimulation and reduced the value of the drive signal to zero. Now, the animal is in a cool environment to which it loses heat at an increased rate. As heat is lost, internal body temperature falls back toward the optimal value. As internal temperature falls, less intense cutaneous cold stimulation is required to balance the effects of the less intense hypothalamic warm stimulation, and so the animal selects a less cool environment to which heat is lost at a less rapid rate. As internal temperature falls to approach the optimal value, the external environmental temperature selected by the animal rises to approach the physiologically neutral value, and the system ends up in a neutral environment with $T_{hy} = T_{hy_0}$, $T_s = T_{s_0}$, and $D = O$.

Thus, the system's responses to external and internal temperature deviations tend to maintain it in the optimal state ($T_{hy} = T_{hy_0}$, $T_s = T_{s_0}$). By adjusting external temperature (toward neutrality in response to an external temperature deviation, and away from neutrality in response to an internal temperature deviation) in such a way as to reduce the absolute value of the drive signal, the system always tends to return to the optimal state following a disturbance. A relatively simple neural system is sufficient to account for these responses, and for the successful temperature regulation which is attendant upon them. The system must operate to minimize the absolute value of the drive signal, and its sensory mechanisms must be "tuned" as indicated above.

Under the conditions of the preference experiments (where the animals were able directly to control internal temperature as well as external temperature), it was possible to demonstrate that the system is able to distinguish internal temperature change from external temperature change, and to minimize the sensory signals as well as the drive signal. Although the results of the preference experiments (Figure 8-4, Table 8-1) demonstrate that these abilities do exist, they are not required (the principle of drive-signal minimization and the assumptions about the "tuning" of the sensory mechanisms are sufficient) to account for successful behavioral temperature regulation under normal conditions (where the animal is able directly to control external temperature, but not internal temperature). The question then arises as to what role is played by these discriminative and sensory signal-minimizing abilities under normal conditions. Presumably, they play a role in governing the selection of particular courses of action from among the various options provided by the environment. For example, a severely hyperthermic animal might choose to enter cool water rather than cool air. Both responses result in drive-signal reduction, but entry into cool water might be preferred because it ensures a more rapid return of internal body temperature to the optimal value.

Comparison of behavioral and automatic systems. It is possible to make some limited comparisons of the information-processing characteristics of the neural systems controlling behavioral and automatic thermoregulatory responses. Hammel and his colleagues (Hammel, 1966; Hammel et al., 1963a; Hammel and Sharp, 1971; Hammel et al., 1963b; Hellstrøm and Hammel, 1967) have made quantitative studies of the interaction of hypothalamic and skin temperatures in the control of several automatic thermoregulatory responses in the dog. One purpose of our studies of the rat (Corbit, 1969, 1970, 1973; Corbit and Ernits, 1971) was to obtain data on the interaction of hypothalamic and skin temperatures in the control of behavioral thermoregulatory responses, which would be comparable with the results of Hammel's studies of the automatic system.

The behavioral and automatic systems are similar in that hypothalamic and skin temperatures have roughly linear effects on both types of response, and in that the same type of additive law (Equation 1) governs the combination of stimuli in both cases. Although both systems make use of hypothalamic and skin temperature information, and both systems apparently combine this information by addition, they may differ in regard to the relative importance of the two stimuli in determining the magnitude of the response. Hammel's studies showed that skin temperature is given less weight than hypothalamic temperature ($b < 1$) in the control of automatic thermoregulatory responses in the dog, whereas we found that skin temperature was given more weight than hypothalamic temperature ($b > 1$) in the control of behavioral thermoregulatory responses in the rat, where b is a dimensionless weighting constant (range, 2 to 3) that expresses the relative importance of skin temperature in determining the rate of responding.

This quantitative difference raises the question of whether the two types of thermoregulatory response are driven by a common drive signal generated by a single integrator, or whether there are separate integrative centers generating separate drive signals for the behavioral and automatic responses. Unfortunately, however, this question can not be satisfactorily answered by comparing results obtained from different species, because the different weighting constants may reflect a difference between rats and dogs rather than between behavioral and automatic systems.

It is possible that there are important species differences in regard to the weighting constant. For example, an a priori case can be made for a relationship between the weighting constant and the body surface-to-mass ratio. A small animal with a high surface-to-mass ratio is seriously threatened by changes in external temperature, and therefore there is a definite adaptive advantage to its giving heavy weight to the signals from its cutaneous thermal receptors. On this basis, one might expect the value of b to be greater in rats than in dogs.

Whether there is a relationship between b and the surface-to-mass ratio, it still may be true that within a given species, skin temperature is weighted more heavily by the behavioral system than by the automatic system. Such a difference would be consistent with the idea that the basic function of the behavioral system is selection of a neutral *external* environment, whereas that of the automatic system is fine-control of *internal* temperature. Our experience with the rat supports this idea to a certain extent. We find that mild warming of the hypothalamus (increases of 1-2° C) elicits full-blown vasodiliation of the skin of the feet and tail in a slightly cool environment (air temperature, 24° C), whereas a 1-2° C change in hypothalamic temperature has a relatively small effect on thermoregulatory behavior (Figures 8-5 and 8-10).

Conclusions

Clearly, we need detailed quantitative studies of the interaction of hypothalamic and skin temperatures in the control of behavioral and automatic thermoregulatory responses in the same species. Such a comparison is being made in the squirrel monkey (Stitt et al., 1971); in that study it is clear that the neutral values of hypothalamic and skin temperatures (T_{hy_0} and T_{s_0}) for the two types of response are quite similar, suggesting that the same pools of central and peripheral receptors and the same sensory information may be used to control the two types of response, but it is too early to tell whether the behavioral and automatic responses are given by the same equation—i.e., whether the values of b are the same.

With knowledge of the laws of the behavioral and the automatic systems in the same species, two important new possibilities will open up. This knowledge will be of value in interpreting the results of electrophysiological studies of the actual neural system. If it should turn out that there are clear quantitative differences between the behavioral and the automatic laws (e.g., different values of the weighting constant, b), then we would have a criterion for deciding whether a given neuron, which is driven both by changes in hypothalamic temperature and by changes in skin temperature (Hellon, 1970), is an element of the behavioral or of the automatic system.

Additionally, increased knowledge of the laws of the behavioral and automatic systems will make it possible to construct models which combine the two systems. With such a synthesis, we begin to approach a realistic conception of the problem of physiological temperature regulation as a whole, and begin to be able to predict thermoregulation by an animal freely interacting with its changing environment.

Supported in part by Research Grant MH 16608 from the National Institutes of Health, U.S. Public Health Service. Thanks are due Russell M. Church, Tiina Ernits, James D. Hardy, Julius W. Kling, and Lorrin A. Riggs for their valuable comments on an earlier draft of this chapter.

References

Adair, E. R., 1971. Displacements of rectal temperature modify behavioral thermoregulation. *Physiol. Behav.* 7:21.

Adair, E. R., Casby, J. U., and Stolwijk, J. A. J., 1970. Behavioral temperature regulation in the squirrel monkey: changes induced by shifts in hypothalamic temperature. *J. Comp. Physiol. Psychol.* 72:17.

Bligh, J., 1966. The thermosensitivity of the hypothalamus and thermoregulation in mammals. *Biol. Rev.* 41:317.

Cabanac, M., 1969. Plaisir ou déplaisir de la sensation thermique et homeothermie. *Physiol. Behav.* 4:359.

Cabanac, M., Duclaux, R., and Gillet, A., 1970. Thermorégulation comportementale chez le chien: effets de la fièvre et de lay thyroxine. *Physiol. Behav.* 5:697.

Carlisle, H. J., 1966. Behavioural significance of hypothalamic temperature-sensitive cells. *Nature* 209:1324.

Chatonnet, J., and Cabanac, M., 1965. The perception of thermal comfort. *Int. J. Biometeor.* 9:183.

Corbit, J. D., 1969. Behavioral regulation of hypothalamic temperature. *Science* 166:256.

Corbit, J. D., 1970. Behavioral regulation of body temperature. In *Physiological and Behavioral Temperature Regulation* (J. D. Hardy, A. P. Gagge, and J. A. J. Stolwijk, eds.). Charles C Thomas, Springfield, Ill., p. 777.

Corbit, J. D., 1973. Voluntary control of hypothalamic temperature. *J. Comp Physiol. Psychol.* 83:394.

Corbit, J. D., and Ernits, T., 1974. Specific preference for hypothalamic cooling. *J. Comp. Physiol. Psychol.* 86:24.

Gagge, A. P., Stolwijk, J. A. J., and Hardy, J. D., 1967. Comfort and thermal sensations and associated physiological responses at various ambient temperatures. *Environmental Res.* 1:1.

Grodins, F. S., 1963. *Control Theory and Biological Systems.* Columbia University Press, New York.

Hammel, H. T., 1966. The regulator of body temperature. *Brody Memorial Lecture VI, Univ. of Missouri Spec. Rep.* 73:1.

Hammel, H. T., 1968. Regulation of internal body temperature. *Ann. Rev. Physiol.* 30:641.

Hammel, H. T., Jackson, D. C., Stolwijk, J. A. J., Hardy, J. D., and Strømme, S. B., 1963a. Temperature regulation by hypothalamic proportional control with an adjustable set point. *J. Appl. Physiol.* 18:1146.

Hammel, H. T., and Sharp, F., 1971. Thermoregulatory salivation in the running dog in response to preoptic heating and cooling. *J. Physiol. (Paris)* 63:260.

Hammel, H. T., Strømme, S. B., and Cornew, R. W., 1963b. Proportionality constant for hypothalamic proportional control of metabolism in unanesthetized dog. *Life Sciences* 2:933.

Hardy, J. D., 1954. Control of heat loss and heat production in physiologic temperature regulation. *The Harvey Lectures 1953-1954, Series XLIX.* Academic Press, New York, p. 242.

Hardy, J. D., 1961. Physiology of temperature regulation. *Physiol. Rev.* 41:521.

Hardy, J. D., 1967. Central and peripheral factors in physiological temperature regulation. In *Les Concepts de Claude Bernard sur le Milieu Interieur.* Masson, Paris, p. 247.

Hardy, J. D., 1969. Brain sensors of temperature. *Brody Memorial Lecture VIII, Univ of Missouri Spec. Rep.* 103:1.

Hellon, R. F., 1970. Hypothalamic neurons responding to changes in hypothalamic and ambient temperatures. In *Physiological and Behavioral Temperature Regulation* (J. D. Hardy, A. P. Gagge, and J. A. J. Stolwijk, eds.). Charles C Thomas, Springfield, Ill., p. 463.

Hellstrøm, B., and Hammel, H. T., 1967. Some characteristics of temperature regulation in the unanesthetized dog. *Am. J. Physiol.* 213:547.

Hensel, H., 1970. Temperature receptors in the skin. In *Physiological and Behavioral Temperature Regulation* (J. D. Hardy, A. P. Gagge, and J. A. J. Stolwijk, eds.). Charles C Thomas, Springfield, Ill., p. 442.

Iggo, A., 1970. The mechanisms of biological temperature reception. In *Physiological and Behavioral Temperature Regulation* (J. D. Hardy, A. P. Gagge, and J. A. J. Stolwijk, eds.). Charles C Thomas, Springfield, Ill., p. 391.

Landgren, S., 1970. The projection of thermoreceptor afferents into the somatosensory system of the cat's brain. In *Physiological and Behavioral Temperature Regulation* (J. D. Hardy, A. P. Gagge, and J. A. J. Stolwijk, eds.). Charles C Thomas, Springfield, Ill., p. 454.

Lipton, J. M., 1971. Behavioral temperature regulation in the rat: effects of thermal stimulation of the medulla. *J. Physiol (Paris)* 63:325.

Matthews, J. T., 1971. Thermal motivation in the rat. *J. Comp. Physiol. Psychol.* 74:240.

Milhorn, H. T., Jr., 1966. *The Application of Control Theory to Physiological Systems.* W. B. Saunders Company, Philadelphia.

Murgatroyd, D., and Hardy, J. D., 1970. Central and peripheral temperatures in behavioral thermoregulation in the rat. In *Physiological and Behavioral Temperature Regulation* (J. D. Hardy, A. P. Gagge, and J. A. J. Stolwijk, eds.). Charles C Thomas, Springfield, Ill., p. 874.

Riggs, D. S., 1963. *The Mathematical Approach to Physiological Problems.* Williams and Wilkins, Baltimore.

Riggs, D. S., 1970. *Control Theory and Physiological Feedback Mechanisms.* Williams and Wilkins, Baltimore.

Satinoff, E., 1964. Behavioral thermoregulation in response to local cooling of the rat brain. *Am. J. Physiol.* 206:1389.

Stevens, J. C., and Stevens, S. S., 1963. The dynamics of subjective warmth and cold. In *Temperature: Its Measurement and Control in Science and Industry,* Vol. 3, Part 3, (J. D. Hardy, ed.), Reinhold, New York, p. 239.

Stitt, J. T., Adair, E. R., Nadel, E. R., and Stolwijk, J. A. J., 1971. The relation between behavior and physiology in the thermoregulatory response of the squirrel monkey. *J. Physiol. (Paris)* 63:424.

Thauer, R., 1964. Der nervöse Mechanismus der chemischen Temperaturregulation des Warmblüters. *Naturwissenschaften* 51:73.

Thauer, R., 1970. Thermosensitivity of the spinal cord. In *Physiological and Behavioral Temperature Regulation* (J. D. Hardy, A. P. Gagge, and J. A. J. Stolwijk, eds.). Charles C Thomas, Springfield, Ill. p, 472.

Weiss, B., and Laties, V. G., 1961. Behavioral thermoregulation. *Science* 133:1338.

Winslow, C.-E. A., Herrington, L. P., and Gagge, A. P., 1937. Relations between atmospheric conditions, physiological reactions and sensations of pleasantness. *Am. J. Hyg.* 26:103.

Yamamoto, W. S., and Brobeck, J. R., eds., 1965. *Physiological Controls and Regulations.* W. B. Saunders Company, Philadelphia.

Zotterman, Y., 1959. Thermal sensations. In *Handbook of Physiology Section I: Neurophysiology,* Volume I. Am. Physiol. Soc., Washington, D. C., p. 431.

9

Beyond the Organism—Social Structure and Energy Metabolism

Edgar B. Hale

The reader of this volume will observe a beautiful flow from the molecular level to the cellular level and on to the organismic level. At that point studies of the control of metabolism come to an abrupt end. This cut-off at the level of the whole animal does not respresent a bias on the part of the editor, but rather reflects the limited insight we biologists have regarding the function of social groups. However, evidence has started to accumulate, especially over the last decade, to support the view that social groups may be treated as energy-trapping systems, with the specific structure of a group representing adaptations to the problems of energy exploitation in a particular ecological situation (Crook, 1965, 1970b,c). This view replaces earlier abortive attempts to explain the evolution of social structures on the basis of phylogeny (Thompson, 1958). The present chapter is intended to provide only an introduction to the problem, not a resolution or even a precise formulation.

Social Groups as Energy-Trapping Systems

General Aspects of Social Structure

For some time biologists have been aware that there were clusters of behavioral and other characteristics associated with specific types of social structure and that a change in one major character was associated with

Table 9-1 Contrasting Characteristics Associated with Two Extreme Types of Social Structure

Behavioral and other characteristics	*Social structure*	
	Breeding pair	*Flock*
Diet	Carnivorous	Herbivorous
Spatial structure	Territorial	Dominance hierarchy
Appearance of sexes	Monomorphic	Dimorphic
Type of mating	Pair-bond	Promiscuous
Development of young	Altricial	Precocial
Care of young	Both sexes	Female only

changes in several functionally related characters (Hinde and Tinbergen, 1958; Hale, 1962a). In other words, certain behavioral characteristics were observed in one type of social structure but not in others. It was also recognized that social structures within the breeding season are usually quite distinct from those during the nonbreeding season (Crook, 1965).

An examination of two extreme types of social structure within the breeding season will illustrate what social structure implies (Table 9-1). The primary ecological aspect associated with these extreme types is that of dietary habit. Associated with the carnivorous diet are a territorial spacing with other breeding pairs excluded from the area, and a morphological similarity of the sexes with adult males and females looking alike and forming a mating bond with a single individual. In addition, the young are unable to move about at birth and must be fed by the parents. There are many variations of social structure between these extremes. But those characteristics listed within columns (Table 9-1) show a high correlation, those between columns, a low correlation.

Correlation Between Habitat and Social Structure

Major impetus for the current investigation of the correlation between contrasts in ecology and social structure came from a comparative study of over fifty species of Weaver birds in Africa by Crook (1962). He found that the type of social structure in a species was closely related to the type of habitat in which the species was evolving (Table 9-2). Crook concluded that the selection pressures responsible for the behavioral characteristics were the type of food, its dispersal throughout the habitat, the skill required in capturing food, and predation pressure. Where considerable skill was required to capture food, the adults were monoga-

Table 9-2. Habitat and Social Structure in 45 Species of Weaver Birds

Habitat	*Social structure*	*Insectivore*		*Grainivore*	
		Monogamous	*Polygamous*	*Monogamous*	*Polygamous*
Forest	Solitary pairs	10	—	—	—
Intermediate	Flocks	2	3?	2	12
Grassy plain	Flocks	—	—	—	16

Adapted from Crook, 1962.

mous and the young were fed by both adults. Crook (1965) subsequently reviewed the correlation between habitat and social structure within the major taxonomic families of birds and reported a general confirmation of his conclusions based on Weaver birds. An extensive review of the ecological adaptations for breeding in birds has been provided by Lack (1968).

In a review of primate social systems, Crook (1970b) concluded that five grades could be defined with reference to ecological factors. Differences in social structures found in forest, savanna, and open country areas were related to seasonal fluctuations in food resources, dispersion of food items, and predation pressure. In some habitats the availability and dispersion of sleeping sites was important.

In a given environment, there is an optimal time and energy budget for a species. A comparison of two species of blackbirds in central coastal California illustrates how the time and energy budgets required for feeding the young may influence the male's role in care of the young (Orians, 1961). During the breeding period the red-winged blackbird occupies large territories and food for the young is gathered either on the territory or adjacent to it. The tricolored blackbird occupies small territories in dense colonies and food for the young is obtained over a large area up to four miles from the nesting site. In both species, the females spend about 15 hours a day feeding the young. In the redwings, with the food supply nearby, this time is adequate and only the female is involved. In the tricolors, with the food supply at a distance, this time is inadequate and both sexes participate in feeding the young.

The number of males associated with the females and young may be related to predation pressure. In the feral sheep on an island in the Hebrides, the rams form a separate social group as is typical of many ungulate species (Grubb and Jewell, 1966). The ewe groups are large, including females of all ages and young males, and consist of matrilinearly related sheep. The ram groups are small and consist of males three years of age or older. Members of this group need not be related. During the breeding season, the ram groups split up and the rams wander about and inseminate the ewes. They play no role in defense of the ewes or lambs.

As predation pressure increases, one or more males may remain with the female groups. In musk oxen, the social group includes several males as well as the females and young. If the group is threatened by a predator, the bulls form a wall against the predator on the threatened side of the herd (Lydeckker, 1898). Similarly, when a troop of baboons is on the move, mothers carrying their infants and the most dominant males are in the center of the group, while the less dominant males and juveniles form the periphery (Hall and DeVore, 1965). If a predator appears, however, the dominant males move to the front to meet the threat. The subordinate males support them, but not too conspicuously. Williams (1966) suggested that in these situations a statistical effect of the differences in behavior of males and females may be at work, with the males not really defending the group. In other words, when confronted by a predator, males face the predator while females and young tend to move away. However, it is more appropriate to consider that the significant adaptation is the incorporation of males into the female group, with the increased effectiveness against predators following as a consequence of that adaptation.

Occasionally, in a given habitat more than one solution to the problems of energy exploitation may evolve, with very different consequences for population numbers. In a comparative study of twenty-four species of Arctic sandpipers, Pitelka (1969) found that some species of these birds have evolved a "conservative" solution. Populations of these species were widely dispersed with a strongly developed territorial system, and there were only minor yearly fluctuations in numbers. Other species had evolved an "opportunistic" solution with dense populations and large yearly fluctuations in population numbers. In the conservative species, the large territories assured adequate food even in years of shortest food supply, and the wide dispersal reduced predation. Mutual care of the young, a persistent pair-bond, and monogamy were characteristics of these species. In the opportunistic species, the territories were much reduced in size during years of high food supply, and the density was much greater. The early departure of one adult reduced the food drain in these circumstances. These species were sexually dimorphic; the males were emancipated from nest activities and mated promiscuously.

These contrasts in group structure have been interspecific. In further studies, more attention may be given to intraspecific variations. Recent studies have revealed intraspecies variations in ungulates (Estes, 1966) and primates (Crook, 1970b) associated with contrasts in ecology. As frequently happens in the history of science, these relationships had been recognized at the start of this century but virtually ignored. While conducting a bibliographic survey, Crook (1970a) discovered that in 1905 Petrucci, a sociologist, had pointed out the marked tendency for similar social structures to emerge when species were living under similar environmental conditions.

Summation

The general picture emerging from these studies is that those aspects of ecology most closely related to the type of social structure which evolves are ones having to do with problems of energy exploitation and retention of energy within the system. The effectiveness of a given social system in exploiting energy resources is associated with the type of food, the distribution and fluctuation of food resources, and the skill required to capture food. Within a social group, the loss of individuals through predation constitutes a major loss of energy to the system. Therefore, optimal retention of energy is closely related to the effectiveness of protection against predators. These observations suggest a view of social groups as energy-trapping systems. Their effectiveness can be measured in terms of their efficiency in exploiting energy resources in a given habitat and in retention of the assimilated energy.

As in other biological systems, social groups are extended in time by reproducing themselves. A type of sexual reproduction occurs in groups involving breeding pairs. A male from one group combines with a female from another group to establish a new social unit. In those species forming larger groups in which an individual spends its entire life within the same group, as in baboon troops or in ewe groups, reproduction takes the form of budding or group fission. Southwick et al. (1965) observed this type of reproduction in a large group of over fifty rhesus monkeys. A subgroup of three males, two females, and six young gradually separated from the larger group, and after a period of several months had become a distinct and independent unit.

This emerging view of social groups as energy-trapping systems may ultimately give us a key to some broader problems in biology. Morowitz (1968), in his book *Energy Flow in Biology,* commented that, although he was able to demonstrate on thermodynamic grounds that ecological systems are ordered energy flow systems, he felt that our analysis of ecology, as well as evolution, was still missing a key. He concluded: "The principle we appear to be missing is a guide as to which of all possible systems will, in fact, arise and evolve in a given energy flow situation." Perhaps the tendency for the appearance of a specific type of social system in a given energy-flow situation offers a clue. Morowitz continued: "There are in biology two levels of order; that concerned with molecular events and that concerned with large-scale events, such as the pattern on butterfly wings or the pecking order in a flock of chickens." Let us note that certain color patterns on the wings of Lepidoptera reduce predation pressure (Blest, 1957). We have noted previously that the peck order (dominance hierarchy) is one of the behavioral characteristics related to social structures in different ecological situations. Is it possible that an increasing efficiency in energy-trapping represents a general parameter, changing in one direction during evolution, from which we might deduce the "creative" basis of evolution?

Animal Culture Under Domestication

In an earlier work I summarized those behavioral characteristics facilitating the domestication of animals (Hale, 1962b). It is noteworthy that most of the characteristics were aspects of social structure or social behavior. In retrospect, it is clear that we domesticate social groups rather than individual animals. Young birds and mammals do not live in social groups by accident but through necessity. They cannot survive outside the social unit without artificial aid. If we remove them from the social group, we must culture them just as we must culture cells or tissues outside the organism. It may be helpful to discuss a few examples of animal culture.

Chick Culture

Under natural conditions, chicks are obligatory members of a social unit consisting of the hen and other chicks. The hen is the control center coordinating the activities of the group, and the chick's information about the environment and its responses to it are monitored through the hen. In addition to maintaining the spatial integrity of the group, the hen plays three roles important in the energetics of the group: she is essential for temperature regulation, location of food and water, and protection against predators. Young chicks are not completely endothermic and require heat from the broody hen to prevent chilling as temperatures drop. Thus the thermoregulatory unit is not a single organism but a social system including the hen and chick. When a chick is chilled, it invokes a specific vocalization for regulating the hen's behavior. Since the hen is constantly losing heat, the energetic cost in regulating the chick's temperature is negligible. The hen also locates food and water for the chicks and gives a specific food call which attracts them to food items. She provides protection against predators by giving an alarm call, hiding the chicks, and confronting the predator.

Under conditions of husbandry, we rear chicks without the hen. Our "chick culturing" is successful only if we fulfill all the functions usually provided by the hen. We accomplish these functions by using a brooder to provide heat, by placing food and water where the chicks cannot possibly miss them, and by building a fence to keep out predators. The fact that cultured chicks and turkeys must learn where food and water are located without help from the mother has led to the myth that young birds must be taught to eat and drink. Similar situations exist for other domestic and laboratory animals.

Hen Culture

Another example of animal culture is illustrated by our culturing of hens in laying houses. Not only are chicks eliminated from this grouping, but for optimal laying, incipient chicks, the eggs, must also be removed at regular intervals. That unselected jungle fowls will lay sixty-two eggs under husbandry conditions, rather than the small number observed in nature, may be attributed to the daily removal of eggs (Hutt, 1949). By permitting eggs to accumulate on the nests, laying turkeys can be induced to start incubating and stop laying (Hale et al., 1969). Similarly, presentation of chicks to hens can inhibit egg laying and induce brooding behavior (Saeki and Tanabe, 1955).

The total dimension of the rooster's role in the chicken flock was discovered recently during a study of the social organization of feral domestic fowl (McBride et al., 1969). During the breeding season he inseminates the females, defends the breeding territory, and then performs other functions we had not anticipated. When a hen is ready to lay she gives a nesting call and moves away from the flock. Soon a male joins her and leads her to nesting sites. The female does not accept the first nesting site shown to her by the male. What is more exasperating, after several sites have been examined, the female sometimes returns to the first site and occupies it. The male then rejoins the flock. After laying, the hen gives the typical cackle which attracts the male. He joins her near the nest and escorts her back to the flock. On the way back they sometimes mate. A continuing problem in hen culture is the large number of eggs laid on the floor rather than in nests. The absence of the male and his escort function may contribute to this problem.

Outside the breeding season, the alpha male is the control center coordinating the activities of the entire flock, and assumes roles for the flock very similar to those played by the hen for her chicks. When a predator appears, he gives an alarm call and hides the group while he confronts the predator. He also maintains the spatial integrity of the flock. Before moving to a new location, he gathers the flock together and then moves off with them. He tidbits at food and calls the flock. While other flock members are feeding, he takes an alert role at the edge of the group. He also inhibits fighting between other members of the flock, thereby promoting group stability.

Some Consequences of Isolation

It is often convenient during experimentation to subject animals to varying degrees of isolation. The consequences of this extreme type of animal culture are in need of intensive study. Rats reared in litter groups and then placed in visual isolation from other rats for several weeks develop abnormal physiological and behavioral symptoms (Hatch et al., 1963). Such animals become difficult to handle, may bite the handler, and show an increased sensitivity to lethal drugs. The behavioral changes were interpreted to be consequences of endocrine changes related to adrenal hyperthrophy. Isolated rats have been found to have circulating levels of corticosterone twice that of grouped animals (R. T. Houlihan, unpublished data). DeLaney (1966) found that isolated rats had a higher mean blood pressure than grouped animals and gave a more extreme response to intra-aortic epinephrine injections. It required at least three months for his animals to adapt to isolation. An aberrant pattern of adrenal secretion was observed in a six-month-old calf which had been maintained in isolation since one day of age (Eberhart, 1966). Responses to ACTH suggested that the calf 's adrenal gland was functioning at near-maximal capacity. If these responses are accepted to be the consequence of prolonged isolation, a tentative possibility subject to further experimentation, we have an example of the inability of a social animal to adapt to conditions of isolation.

Behavior not previously present may occur under conditions of isolation. In a strain of rats in which only 4% were mouse-killers, visual isolation for a period of fourteen weeks increased the incidence to over 50% (Pion, 1970). Even though behavioral systems may develop fully, animals reared in isolation fail to achieve an integrated balance between systems, such as those monitoring fighting and sexual behavior, and overrespond in a way that precludes effective social interactions (Kruijt, 1964; Harlow et al., 1966).

Regulation and Control Within Social Groups

In many instances, certain individuals in the group serve as control centers, coordinating the activities of the group. These individuals monitor temporal changes in the environment and, through a set of mutual signals with other group members, bring about appropriate adjustments within the group. It may be helpful to consider a few examples of temporal coordination between members of the social group. To the extent that the behavior of an individual animal contributes to changes and regulation within the social unit, we may think of the behavior and its consequences as representing the "physiology" of the social system.

Transfer of Energy

The energetic efficiency of a social group is greater than that of an individual organism if losses to predation are reduced to zero. This increased efficiency accrues from the utilization of waste heat and from the transfer of food materials within the social group. As we noted in the case of the chick, and to an even greater extent in altricial young, waste heat from the mother contributes to the maintenance of high body temperature during the early ectothermic stages. Consequently, a larger percentage of the energy from their food is available for growth (Bartholomew, 1968). Transfer of energy within the group is accomplished by lactation in mammals. In many species of birds, the adults regurgitate food for the young (Lehrman, 1961). An adult can bring food to the young at less energy cost than that required for the young, with a higher metabolic rate, to obtain food directly. Even if the parent does no more than locate food and water for the young, less energy is required than would be expended by the inefficient searching of inexperienced young.

One class of individuals within the social group may modify the metabolic activity of another class. A drastic shift in metabolic pattern occurs when birds are engaged in incubation activities. In many species, this shift includes an inhibition of food intake. A turkey hen incubating artificial eggs may continue to incubate until she dies of starvation (W. M. Schleidt, personal communication). Fortunately, fertile eggs hatch after twenty-seven days, the inhibitory effect of the egg is thereby removed, and the hen's metabolic activity and regulation of food intake revert to the preincubation pattern.

Endocrine Control

As indicated earlier, a few studies suggest that aberrant endocrine responses develop in animals separated from members of their own species. That the presence or behavior of one class of animals in the social group can modify the endocrine activity of another class has been thoroughly documented (Lehrman, 1961). This type of regulation promotes temporal synchronization of endocrine and behavioral functions between different types of individuals in the social group. As noted, chicks inhibit ovulation and promote broody behavior in the hen. Once the chicks become too large to constitute a "chick" stimulus, the domesticated hen ignores them and starts ovulating again. Thibault et al. (1966) concluded that the delayed return to estrus in suckling ewes, relative to milking ewes, indicated that stimuli from the lamb inhibited sexual activity of the mother.

Following the now-classical observation that the female pigeon could be induced to lay eggs by seeing a male, numerous studies have demonstrated that the presence and behavior of the male may stimulate the secretion of gonadotrophic hormones in female birds (Lehrman, 1961). In some instances, the vocalizations of the male may stimulate ovarian activity. This type of synchronization of endocrine activity and behavior in the two sexes is especially evident in species that form pair-bonds with both parents participating in care of the young. In species that breed in flocks with promiscuous matings, this type of synchronization may not occur. For example, the onset of ovulation in turkey females is not influenced by the presence or absence of courting males (Hale et al., 1969).

In domestic mammals, it has been found that the introduction of the ram stimulates ovulation and subsequent estrus in ewes during that time of the year when the ewes are not cycling normally (Hulet, 1966). This effect was produced in ewes which could either see, smell, or hear the rams. While olfactory stimulation from the male will accelerate estrus in female mice, exposure to the odor of a strange male blocks pregnancy in newly mated females (Bruce, 1966). Removal of the olfactory bulbs in sows either completely eliminates estrus or produces irregular cycles (Thibault et al., 1966).

Behavioral Control

Exchange of signals regulating behavior between members of a social group must be essentially error-free to avoid misspent effort and waste of energy. Failure to provide appropriate signals may sometimes be a matter of life or death. The peeping of the young turkey poult not only attracts the attention of the hen but also protects it from attack by the hen (Hale et al., 1969). Deafened hens attack and kill their own young immediately after the young are hatched. In addition, normal hens presented with silent dummy poults attack them, while the same dummy poults emitting recorded peeps are not attacked. In rats, the pups of females that kill mice are protected by their odors from attack by their own mothers. If the odor of the pups is disguised or eliminated, the female attacks and kills them (Myer, 1964).

Failure to distinguish between "operational error" and "potential error" in signaling systems has led to misinterpretations about the efficiency of such systems. If an animal's behavior is to lead to reliable temporal adjustments synchronizing its behavior with changes in the environment, the significant aspects of the environment must be coded. The animal's behavior is then mediated efficiently by matching the environmental code with an internal model of the environment. Usually the coding of the environment is detailed only to the extent required to minimize operational error in the specific environment in which the species is evolving. Consequently, in the natural habitat the behavior operates reliably with low error. Under artificial conditions, however, there may be items—many or few—which match the code. In the first instance, the potential error is very high, with the animal showing inappropriate behavior toward those items. In the second instance, the potential error as well as the operational error remains low.

In domestic animals, the potential error in a code may be used advantageously or it may create problems. In bulls, the code for a receptive female is very imprecise and anything that can be mounted and does not move away matches the internal model (Hale, 1966). However, the operational error in this code is very low since a bull spending his life in a normal social group encounters only one thing which matches the internal model—a cow in estrus. The young bull quickly learns not to mount those animals that move away. In contrast, the potential error is very great if the bull is removed from his natural habitat. In a laboratory, there may be many inanimate objects and animals which match the code. In breeding studs, we take advantage of this potential error by using restrained anestrous cows, restrained bulls, or even dummies to stimulate bulls for semen collection. The potential error provided by certain sounds present under husbandry conditions may have disastrous consequences. Young turkeys respond to a particular alarm call given by the mother by running and crouching near some object (Hale et al., 1969). Unfortunately, some of the sounds made by pit cleaners match this call and cause the young birds to pile in the corners of the pen where they may smother if not detected in time.

Among individuals of the same class (e.g., males or females), the primary behavioral interactions are related to spatial adjustments. By maintaining appropriate individual distances, the animals keep aggressive interactions at a minimum and conserve energy otherwise wasted in fighting (McBride, 1964). The patterns of facing behavior in adult hens—

the distances maintained between individuals—is most revealing (McBride et al., 1963; McBride, 1964). Hens more than 30 inches apart are relatively unconcerned with each other; between 16 and 30 inches they turn their heads to avoid each other's faces; but if they approach within 16 inches, they face beak to beak in an aggressive or defensive posture. Chance (1962) has suggested that "cut-off" acts or postures, similar to the "facing away" in chickens, provide a common behavioral mode reducing aggressive interactions in animals. In young primates, it is important to distinguish between play-fighting and serious aggressive behavior. If a young monkey wants to play-fight, he first gives a signal indicating that he is playing. His attack is then responded to as play (Altmann, 1962). Here we have an example of a signal which indicates the quality of the communication that is going to follow.

Breakdown in Social Structure

Under conditions of high population density, the behavioral patterns regulating spatial adjustments may fail, with a consequent breakdown in social structure. In rodent populations, increased aggressive behavior erupts under conditions of overcrowding, and is related to the inability of the animals to maintain their social structure (Archer, 1970). A possible interpretation is that at high densities the various cut-off postures effective at lower densities can no longer function, and the observed aggressive behavior is symptomatic of the breakdown in social structure. Chickens housed at high densities sometimes develop an abnormal behavior called avian hysteria. The birds face in the same direction and run in a stream about the house for several minutes. After a brief break, the running is resumed and this pattern is repeated throughout the daylight hours to end only with the onset of darkness. Egg laying is inhibited and a serious economic loss is incurred. The genesis of this behavior may be the inability of a bird to face away from the nearest bird without coming face to face with another one. Juvenile chickens sometimes engage briefly in a similar streaming run as isolated broods join to form a single flock (McBride et al., 1969). Reverting to this type of streaming behavior under high densities permits the birds to all face in the same direction and gain temporary relief from the bind of not being able to face away.

Removal or disintegration of the controls within a social system may lead to dire consequences. Readers familiar with Golding's *Lord of the Flies* may recall the behavior of a group of young boys shipwrecked on an island and removed from the control of adults. Most reviewers attributed the subsequent behavior to the inherent evil in man. I do not know what Golding had in mind, but he depicted precisely what happens to any organized system when the controls are removed—it turns into a monstrosity.

Summary and Conclusions

Evidence is accumulating which demonstrates a high correlation between social structures and the type of habitat in which a species is evolving. Those aspects of ecology most closely related to the type of social structure which evolves are ones having to do with problems of energy exploitation and retention of energy within the social system. As is true of other biological systems, social groups are extended in time by reproducing themselves.

The perspective provided by this view of social structures as energy-trapping systems has implications for problems of animal culture under domestication. We domesticate animal groups rather than individual animals. As we remove animals from the natural social group, we must "culture" them just as we culture cells or tissues outside the organism. Our culturing is successful only to the extent that our husbandry provides all those functions operative in the natural group. Some animals may be unable to adapt to complete isolation from members of their own species, and develop unstable endocrine responses.

In natural groups, certain individuals serve as control centers, monitor changes in the environment, and, through communication signals, bring about appropriate adjustments within the group. The energetic efficiency of a social group is higher than that of individual organisms if losses to predation are reduced to zero. This increased efficiency accrues from the utilization of waste heat and the transfer of food material within the group, and continued normal endocrine levels and thus, less irrational, energy-wasting behavior.

One class of individuals (e.g., males or females) in the social group may modify the endocrine functions of another class. This type of regulation promotes the temporal synchronization of behavioral functions between different classes within the social unit. The exchange of signals regulating behavior within the group must be essentially error-free to avoid misspent effort and consequent waste of energy. However, in work with domestic or laboratory animals, the potential error in a signaling system may sometimes be used advantageously. A breakdown in social structure may lead to increased aggressive behavior with energy channeled into nonproductive activities.

References

Altmann, S., 1962. A field study of the sociobiology of rhesus monkeys, *Macaca mulatta. Ann. N.Y. Acad. Sci.* 102:338.

Archer, J., 1970. Effects of population density on behavior in rodents. In *Social Behavior in Birds and Mammals* (J. H. Crook, ed.). Academic Press, New York, p. 169.

Bartholomew, G. A., 1968. Body temperature and energy metabolism. In *Animal Function: Principles and Adaptations* (M. S. Gordon, ed.). The Macmillan Company, New York, p. 290.

Blest, A. D., 1957. The evolution of protective displays in the Saturnioidea and Sphingidae (Lepidoptera). *Behaviour* 11:257.

Bruce, H. M., 1966. Smell as an exteroceptive factor. *J. Anim. Sci.* (Suppl.) 25:83.

Chance, M. R. A., 1962. An interpretation of some agonistic postures: the role of "cut-off" acts and postures. *Symp. Zool. Soc. Lond.* 8:71.

Crook, J. H., 1962. The adaptive significance of pair formation types in weaver birds. *Symp. Zool. Soc. Lond.* 8:57.

Crook, J. H., 1965. The adaptive significance of avian social organizations. *Symp. Zool. Soc. Lond.* 14:181.

Crook, J. H., 1970a. Social behavior and ethology. In *Social Behavior in Birds and Mammals* (J. H. Crook, ed.). Academic Press, New York, p. xxi.

Crook, J. H., 1970b. The socio-ecology of primates. In *Social Behavior in Birds and Mammals* (J. H. Crook, ed.). Academic Press, New York, p. 103.

Crook, J. H., 1970c. Social organization and the environment: Aspects of contemporary social ethology. *Anim. Behav.* 18:197.

DeLaney, R. G., 1966. The effect of long-term isolation on blood pressure, behavior, and cardiovascular response to large doses of epinephrine in the rat. Master's thesis. The Pennsylvania State University.

Eberhart, R. J., 1966. Glucocorticoid secretion in cattle. Ph.D. diss. The Pennsylvania State University.

Estes, R. D., 1966. Behavior and life history of the Wildebeest *(Connochaetes taurinus). Nature* 212:999.

Grubb, P., and Jewell, P. A., 1966. Social grouping and home range in feral soay sheep. *Symp. Zool. Soc. Lond.* 18:179.

Hale, E. B., 1962a. Comparative behavior. In *The Behavior of Domestic Animals* (E. S. E. Hafez, ed.). Williams and Wilkins, Baltimore, p. 589.

Hale, E. B., 1962b. Domestication and the evolution of behavior. In *The Behavior of Domestic Animals* (E. S. E. Hafez, ed.). Williams and Wilkins, Baltimore, p. 21.

Hale, E. B., 1966. Visual stimuli and reproductive behavior in bulls. *J. Anim. Sci.* Suppl. 25:36.

Hale, E. B., Schleidt, W. M., and Schein, M. W., 1969. The behavior of turkeys. In *The Behavior of Domestic Animals,* (ed. 2) (E. S. E. Hafez, ed.). Williams and Wilkins, Baltimore, p. 554.

Hall, K. R. L., and DeVore, I., 1965. Baboon social behavior. In *Primate Behavior: Field Studies of Monkeys and Apes* (I. DeVore, ed.). Holt, Rinehart and Winston, New York, p. 53.

Harlow, H. F., Joslyn, W. D., Senko, M. G., and Dopp, A., 1966. Behavioral aspects of reproduction in primates. *J. Anim. Sci.* (Suppl.) 25:49.

Hatch, A., Balazs, T., Wiberg, G. S., and Grice, H. C., 1963. Long-term isolation stress in rats. *Science* 142:507.

Hinde, R. A., and Tinbergen, N., 1958. The comparative study of species-specific behavior. In *Behavior and Evolution* (A. Roe and G. G. Simpson, eds.). Yale University Press, New Haven, p. 251.

Hulet, C. V., 1966. Behavioral, social and psychological factors affecting mating time and breeding efficiency in sheep. *J. Anim. Sci.* (Suppl.) 25:5.

Hutt, F. B., 1949. *Genetics of the Fowl.* McGraw-Hill, New York, p. 287.

Kruijt, J. P., 1964. Ontogeny of social behavior in Burmese Red Jungle fowl *(Gallus gallus spadiceus). Behaviour* (Suppl. 12).

Lack, D., 1968. *Ecological Adaptations for Breeding in Birds.* Methuen, London.

Lehrman, D. S., 1961. Hormonal regulation of parental behavior in birds and infrahuman mammals. In *Sex and Internal Secretions,* Vol. 2 (W. C. Young, ed.). Williams and Wilkins, Baltimore, p. 1268.

Lydeckker, R., 1898. *Wild Oxen, Sheep, and Goats of All Lands.* Rowland Ward, London.

McBride, G., 1964. A general theory of social organization and behavior. *Univ. Queensland Papers, Vet. Sci.* 1:75.

McBride, G., James, J. W., and Shoffner, R. N., 1963. Social forces determining spacing and head orientation in a flock of domestic hens. *Nature* 197:1272.

McBride, G., Parer, I. P., and Foenander, F., 1969. The social organization of feral domestic fowl. *Anim. Behav. Monogr.* 2:127.

Morowitz, H. J., 1968. *Energy Flow in Biology.* Academic Press, New York, p. 120, 137.

Myer, J. S., 1964. Stimulus control of mouse-killing rats. *J. Comp. Physiol. Psychol.* 58:112.

Orians, G. H., 1961. The ecology of blackbird *(Agelaius)* social systems. *Ecol. Monogr.* 31:285.

Pion, L. V., 1970. Effects of experience, social structure and hunger on the incidence of mouse-killing behavior in Norway rats. Ph.D. diss. The Pennsylvania State University.

Pitelka, F. A., 1969. Evolution of social organization in Artic sandpiper populations. *Proc. XI Int. Etholog. Congr.* p. 75.

Saeki, Y., and Tanabe, Y., 1955. Changes in prolactin content of fowl pituitary during broody periods and some experiments on the induction of broodiness. *Poultry Sci.* 34:909.

Southwick, C. H., Beg, M. A., and Siddiqi, M. R., 1965. Rhesus monkeys in North India. In *Primate Behavior: Field Studies of Monkeys and Apes* (I. DeVore, ed.). Holt, Rinehart and Winston, New York, p. 111.

Thibault, C., Courot, M., Martinet, L., Mauleon, P., du Mesnil du Buisson, F., Ortavant, R., Pelletier, J., and Signoret, J. P., 1966. Regulation of breeding season and estrous cycles by light and external stimuli in some mammals. *J. Anim. Sci.* (Suppl.) 25:119.

Thompson, W. R., 1958. Social behavior. In *Behavior and Evolution* (A. Roe and G. G. Simpson, eds.). Yale University Press, New Haven, p. 291.

Williams, G. C., 1966. *Adaptation and Natural Selection.* Princeton University Press, Princeton, p. 219.

10

The Special Challenge of Biological Control Systems

Howard T. Milhorn, Jr.

In investigating biological control systems, we use, in a loose sense, what the engineer calls systems analysis. The engineer generally knows all of the components of the system with which he is working, and he knows or can readily determine the mathematical expression of each. He also knows how these components are interconnected and where feedback loops exist. He is interested in the response of the system as a matter of design. The biological investigator, on the other hand, is not concerned with design. The system in which he is interested has already been designed. He is challenged with the difficult task of determining, by experimentation, the mathematical or graphical expressions of the components of biological systems, and identifying their interconnections. Only when this major task is accomplished can he begin to understand how the biological system functions as a whole.

Some of the techniques of control theory offer methods for determining the mathematical expressions of the components of biological systems. Computer simulation offers a method for investigating in detail how these components work together to meet the needs and desires of the organism.

Disease states, in many instances, are merely malfunctions of biological control systems. That medicine too can benefit from a knowledge of systems analysis is pointed out only too well by Norbert Wiener's (1961) classical example: A patient comes into a neurological clinic. He is not paralyzed and he can move both legs when given the order. Yet he walks with an uncertain gait, looking at the ground. He starts each step with a kick, throwing each foot in front of him in succession. If blindfolded, he cannot stand and totters to the ground. This patient suffers from an abnormality of a particular control system. He has *tabes dorsalis*. The part of his spinal cord that receives information has been damaged or destroyed by an infectious disease. His brain is unable to tell the position and state of motion of his feet. In other words, he has lost part of his proprioceptive sense. He must rely upon his eyes and the balancing organs of his inner ear as a substitute. A secondary feedback loop has, therefore, been substituted for the primary one. When he is blindfolded, the secondary loop is broken and he is operating in the open-loop state.

Perhaps the most important thing that comes from the application of systems analysis techniques to biological systems is the interest it stimulates. The investigator begins to look at many systems in a new light. He begins to recognize where the gaps exist in his knowledge of a particular system; in order to understand in detail how a system works, he must have a complete system with which to work. The components of this system must be clearly identified and the feedback loops must be known or determined.

One danger must be recognized in applying systems analysis techniques to biological systems. There are basic dissimilarities between technological and biological systems. Therefore, in applying systems analysis techniques to biological systems the investigator should be certain that he is doing just this and not the opposite—i.e., attempting to apply biological systems to systems analysis techniques.

Here are a few of the different ways systems analysis has been applied. These examples are not to be considered all-inclusive or even the best of their kind. They are merely the examples with which I am most familiar.

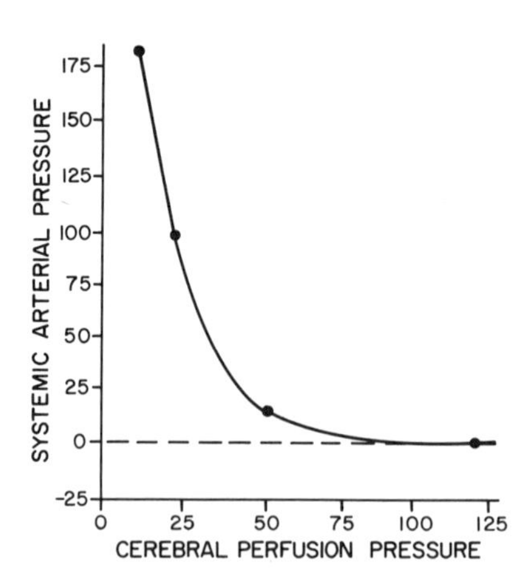

Figure 10-1. Experimental relationship between cerebral perfusion pressure (CPP) and the change in systemic arterial pressure (SAP). (After Sagawa et al., 1961a)

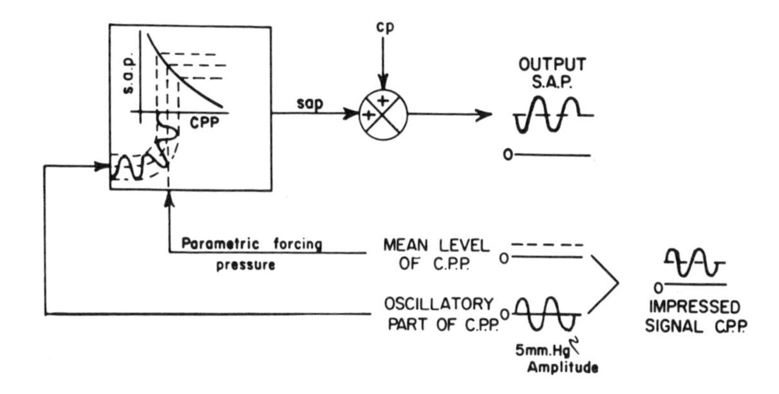

Figure 10-2. Frequency analysis of the CNS ischemic response. (After Sagawa et al., 1961b)

Frequency Response

Frequency response is a tool that is useful to the control engineer. By imposing constant-amplitude signals of different frequencies on a system, he is able to describe the characteristics of the system and even to derive a mathematic expression of its behavior (the transfer function). Sagawa et al. (1961a,b) used this method to investigate the stability of the central nervous system (CNS) ischemic response. It was proposed that vasomotor waves in arterial pressure were due to feedback between the CNS and the systemic circulation. For this study, the heads of dogs were isolated from their bodies with the exception of the nervous system. The systemic arterial pressure (SAP) response to changes in cerebral perfusion pressure (CPP) was then investigated for linearity (Figure 10-1). Since the relationship is clearly nonlinear, it was handled in a special way. A pressure of interest was chosen and the frequency input around this pressure was kept at a small amplitude to approximate linearity, as illustrated in Figure 10-2. The impressed signal is composed of the chosen mean level and the oscillatory part. The output, the systemic arterial pressure, is then recorded for several different frequencies, three of which are shown in Figure 10-3. Next, the system was investigated for transportation lag. A sudden step in CPP was introduced, as shown in Figure 10-4. After a lag (L), SAP began to increase. Using the data obtained from the frequency inputs and the transportation lag, Bode plots were drawn (Figure 10-5). The upper curve relates the gain of the system in decibels to the input frequency. The lower curve relates the

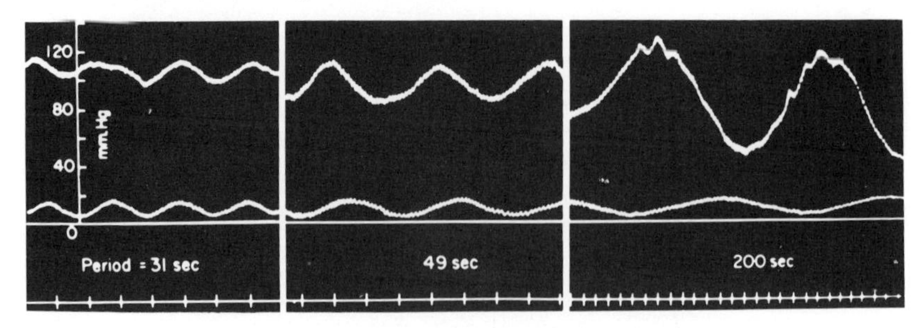

Figure 10-3. Response of the CNS ischemic response to three sinusoids of different frequencies. (Upper trace is the output; lower trace, the input.) (Sagawa et al., 1961b)

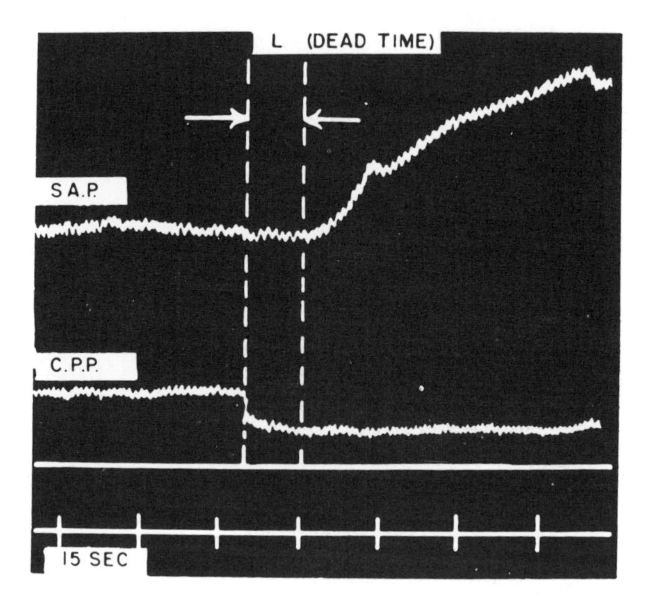

Figure 10-4. Transportation lag of the CNS ischemic response. Dead time, L, determination in transient ischemic response. (Sagawa et al., 1961b)

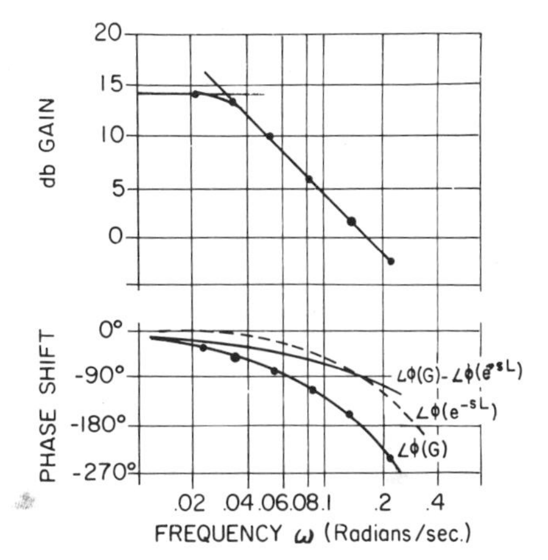

Figure 10-5. Bode plots of the CNS ischemic reflex response. See text for explanation. (After Sagawa et al., 1961b)

phase shift between input and output to the input frequencies. The points along the solid line are the phase data from the input frequencies. The transportation lag (broken line) and the exponential lag of the system are its components. For this particular experiment, by a systems analysis method a mathematical expression for the response was determined:

$$\text{SAP} = [e^{-sL}/(\tau s + 1)]\ \text{CPP}$$

From Bode plots, Nyquist diagrams were drawn to show the stability of the system at a glance. If a plot encircles the -1 point, the system is unstable. If it passes inside of the -1 point, the system is stable. Figure 10-6 shows Nyquist plots of the CNS ischemic response for mean pressures of 60, 40, and 30 mm Hg. As can be seen, for this particular animal, instability does not result until the pressure is reduced to 30 mm Hg. From this study, it was concluded that vasomotor waves following a drastic reduction in arterial pressure do result from a CNS-to-peripheral-circulation feedback.

Simulation as an Investigative Tool

As an example, let us consider work on hypertension by Guyton and Coleman (1969). Because of the many factors involved in the regulation

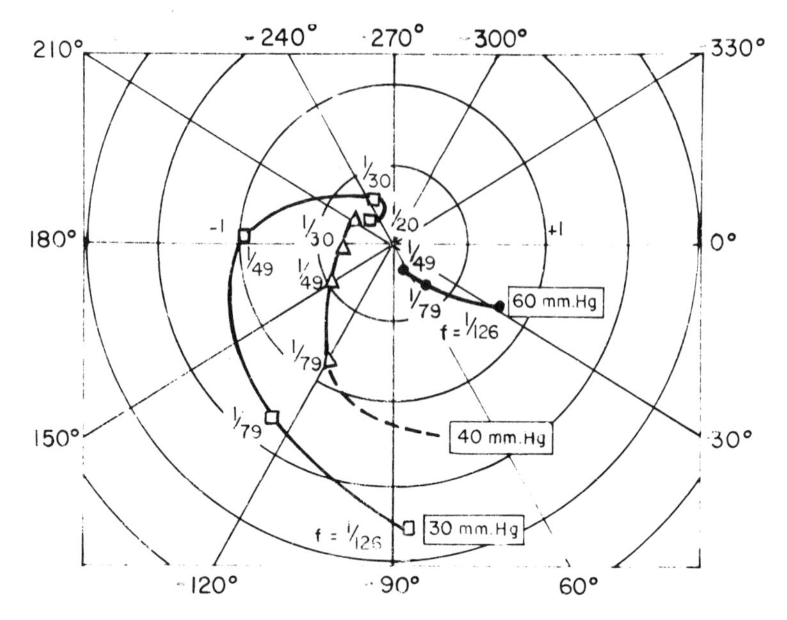

Figure 10-6. Nyquist plot of the CNS ischemic reflex response. (Sagawa et al., 1961b)

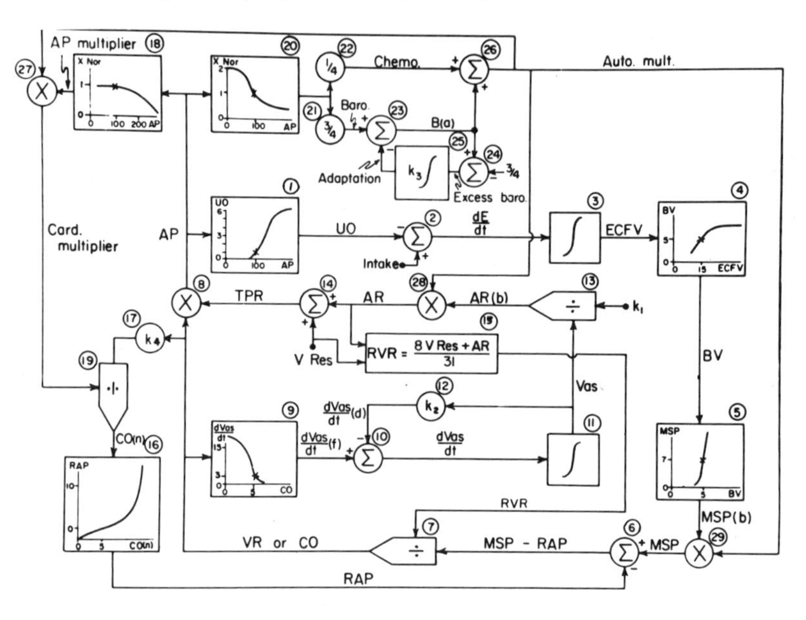

Figure 10-7. Block diagram of long-term circulatory control. (Guyton and Coleman, 1969)

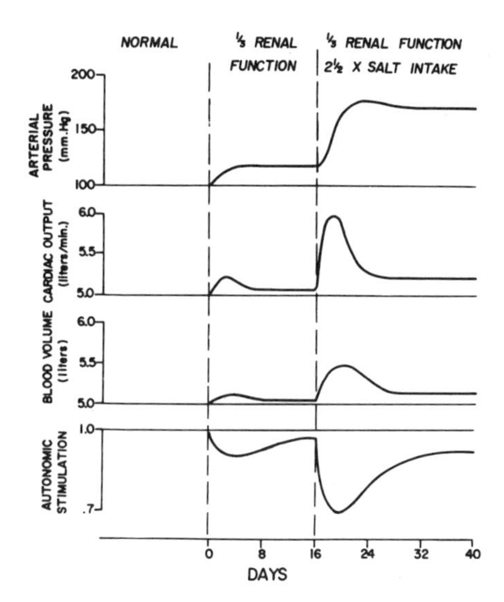

Figure 10-8. Responses from the model to reduction of renal mass and to salt loading. (Guyton and Coleman, 1969)

of arterial pressure, it is impossible to visualize quantitatively what is happening during various disturbances to the system. Therefore, computer simulation was used. The basic model is shown in Figure 10-7. Because of its complexity, we will not go into it here. The interested reader can refer to the original source.

Responses from the model for salt-induced hypertension are shown in Figure 10-8. At time zero, the renal mass is reduced to one-third of its normal value, simulating reduced kidney function. Note that only mild responses occur in all variables. However, when the salt intake is increased to two and one-half times normal, drastic changes occur. Note that arterial pressure rises slowly and remains elevated. Cardiac output, however, rises rapidly, peaks, and then returns to a low but still increased level. Blood volume also shows a similar transient increase. Autonomic activity decreases rapidly, undershoots, and then rises back to a level still below the control value. It should be noted that arterial pressure rises more slowly than cardiac output or blood volume and that it continues to rise even after cardiac output has peaked and is falling back toward a new level. This is because as long as there is excess blood flow through the tissues, the phenomenon of autoregulation continues to increase the peripheral resistance in an attempt to return the flow to normal. A second reason for the continued rise of arterial pressure is baroreceptor adaptation.

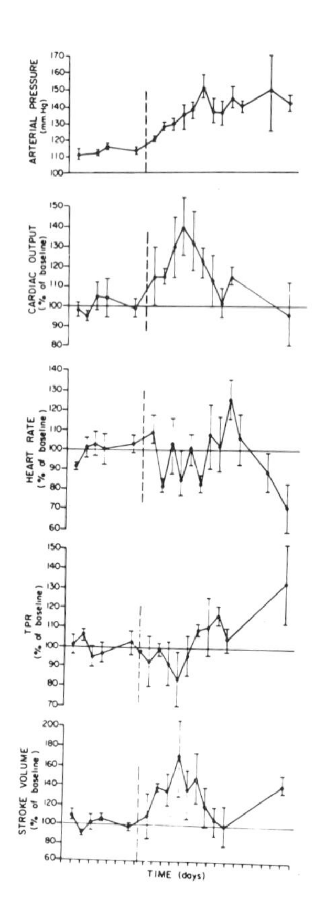

Figure 10-9. Experimental verification of the predictions of the model in Fig. 10-8. (Coleman and Guyton, 1969)

Experimental verification of the predictions of the model is shown in Figure 10-9. As can be seen, arterial pressure and cardiac output exhibit the same type of response as the model. In addition, heart rate, as an indication of sympathetic activity, decreases as predicted for autonomic stimulation. Total peripheral resistance (TPR) increases as predicted for blood volume, to cause the decline in cardiac output from its maximum value.

From the analysis, investigators noted some interesting points about this model: (1) autoregulation is an important factor in hypertension, and (2) almost no effect can cause continued elevation of arterial pressure unless the kidneys are involved.

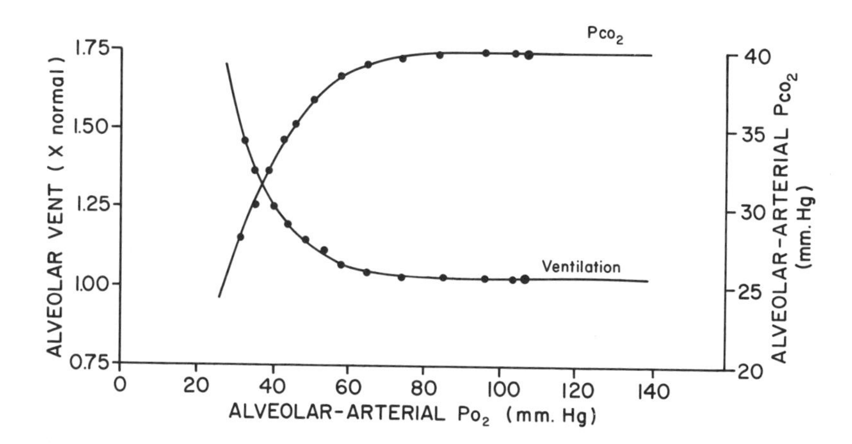

Figure 10-10. Steady-state relationship between P_{AO_2}, P_{AO_3}, and ventilation. (Gray, 1950)

The Use of External Control Systems

Often, the biological relationship we wish to investigate is more involved than a simple relationship between two variables. Three or more variables are generally encountered. In some cases, it is possible to reduce a multivariable system to a two-variable system with the use of one or more external control systems (Milhorn, 1966, 1969). Figure 10-10 illustrates the responses of a three-variable system, the respiratory system, to varying degrees of hypoxia. As alveolar P_{O_2} is lowered, ventilation increases and causes alveolar P_{CO_2} to fall. It is well known that low P_{CO_2} acts as an inhibitor of ventilation. Suppose we wish to study the relationship between alveolar P_{O_2} and ventilation independent of alveolar P_{CO_2}. We would need an external control system. Such a system was designed by Holloman et al. (1968). Shown in Figure 10-11, it operates as follows: the subject breathes from a mask to which a pneumotach is connected for use with a transducer to monitor airflow. Carbon dioxide is monitored continuously by an infrared analyzer and linearizer. At the end of each expiration expired CO_2 is sampled by a trigger circuit that operates on the airflow signal. It is held at that value until the next sample is taken by a zero-order hold circuit. The output is the actual alveolar (end-tidal) CO_2 It is compared to a preset reference

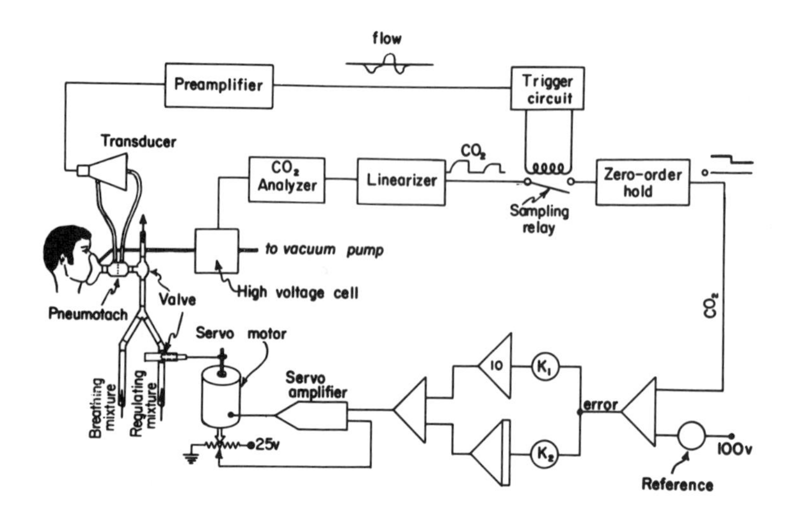

Figure 10-11. A sampled-data regulator for maintaining alveolar P_{CO_2} constant during hypoxia. (Holloman et al., 1968)

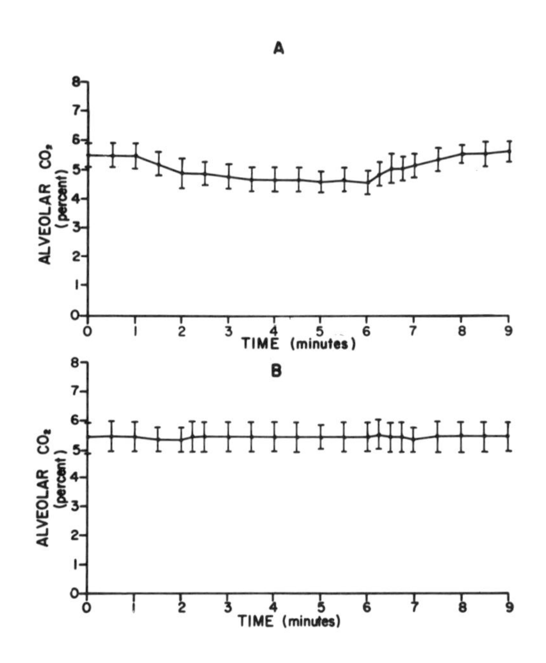

Figure 10-12. Response of end-tidal CO_2 to hypoxia. A, without controlled CO_2; B, with controlled CO_2. (Holloman and Milhorn, unpublished data)

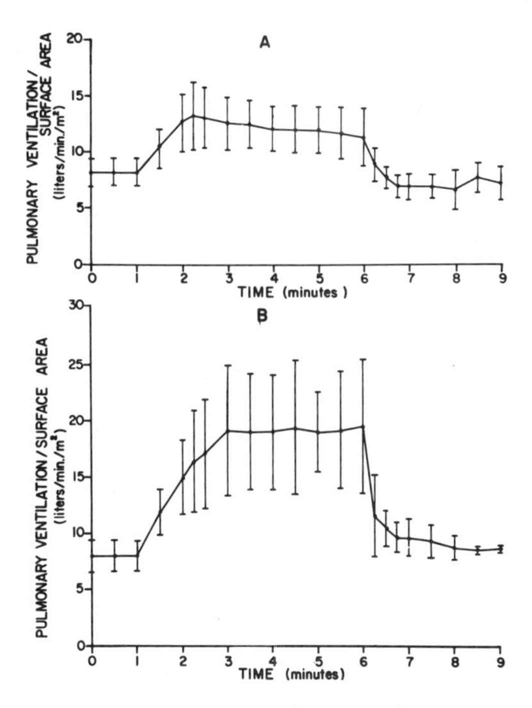

Figure 10-13. Response of minute-by-minute ventilation to hypoxia. A, without controlled CO_2; B, with controlled CO_2. (Holloman and Milhorn, unpublished data)

to give an error signal. By use of proportional plus integral control, a position-servo drives a valve. When the actual CO_2 is less than the reference setting the valve automatically opens, allowing a high concentration of CO_2 (regulating mixture) to flow into the breathing mixture until the error is removed. On the other hand, when the actual CO_2 is greater than the reference setting, the valve closes until the error is removed.

When a subject is allowed to breathe a mixture low in oxygen, ventilation increases; as a result the CO_2 begins to fall. The external control system immediately opens the valve to allow the high-CO_2 mixture to enter the breathing mixture as needed to maintain end-tidal CO_2 at a constant value. Figure 10-12 shows the response of alveolar CO_2 when human subjects were allowed to breathe 9% oxygen in a step fashion without controlled CO_2 (A, at top) and with controlled CO_2 (B, at bottom). Figure 10-13 shows the resulting minute-by-minute ventilation without controlled CO_2 (A, at top) and with controlled CO_2 (B, at bottom). The difference in the two records represents ventilatory inhibition by lowered CO_2.

Positive Feedback

The systems discussed above operate by negative feedback. Positive feedback, however, does exist in biological systems. Circulatory shock may be one such situation. Studies by Guyton and Crowell (1964) indicate that the heart deteriorates severely as shock progresses. Once the severity has reached a certain level, it becomes progressively worse until death occurs. The positive feedback loops thought responsible (Guyton and Crowell, 1964) are shown below.

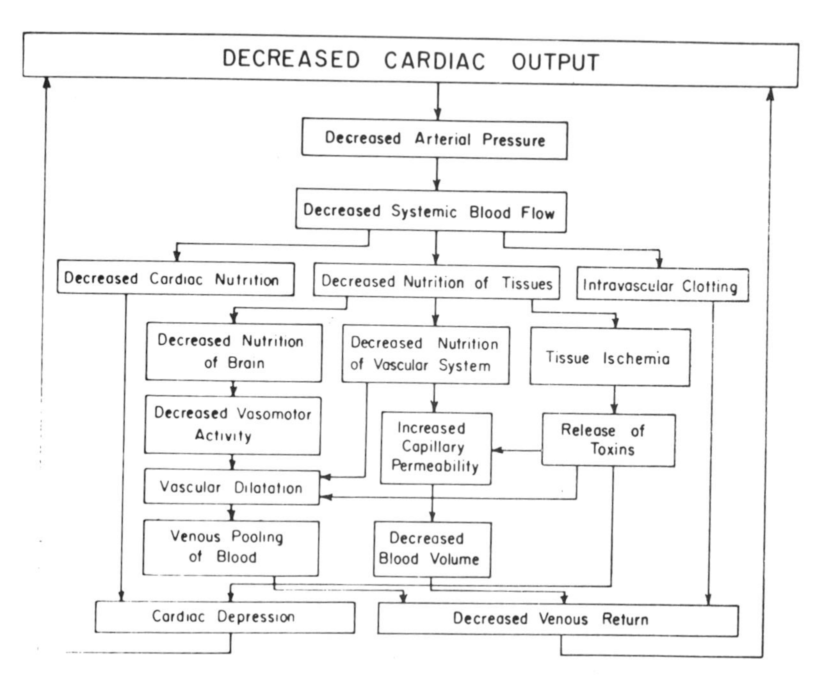

Any factor that sufficiently decreases cardiac output, such as hemorrhage, is capable of throwing the system into shock as follows: the decreased cardiac output decreases the arterial pressure which, in turn, decreases the systemic blood flow. Several things then happen, including decreased cardiac nutrition, decreased nutrition of other tissues, and possibly intravascular clotting. The end result is cardiac depression and decreased venous return. Both of these further decrease the cardiac output. An initial decrease in cardiac output has resulted in a further decrease; hence, positive feedback.

Small initial decreases in cardiac output do not normally lead to circulatory shock because the positive feedback is opposed by stronger nega-

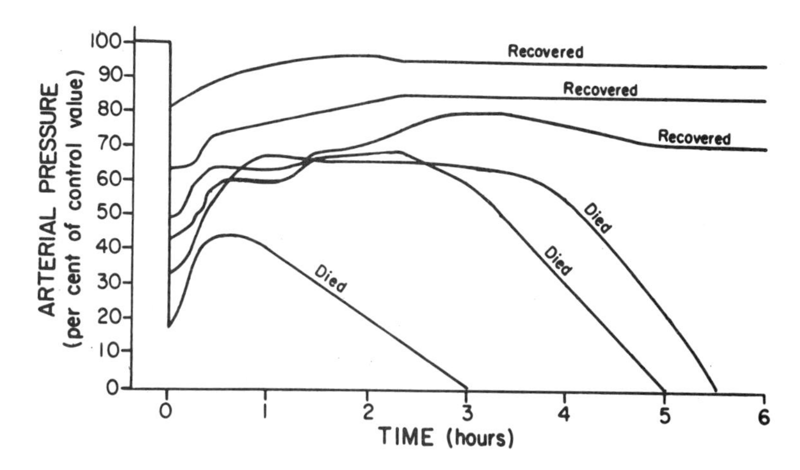

Figure 10-14. Course of arterial pressure after different degrees of hemorrhage. (Guyton and Crowell, 1961)

tive feedback. A point is reached, however, when the cardiac output is decreased far enough that the positive feedback becomes stronger than the negative feedback, thus leading to progressive shock. This is illustrated in Figure 10-14. When arterial pressure of dogs is decreased slightly by bleeding, recovery occurs rapidly. Severe bleeding, however, results in death. When the pressure is decreased to intermediate levels, the animals lived for several hours with death or recovery uncertain until, finally, positive or negative feedback gained control and the animals then went on to die or recover.

Conclusions

The four examples I have presented are, of course, only a few of the ways in which systems analysis has been used by the biological investigator. The key word describing each example is *quantitation*. Systems analysis offers a special challenge to the investigator interested in a useful tool for quantitating biological phenomena at the systems level.

References

Coleman, T. G., and Guyton, A. C., 1969. Hypertension caused by salt loading in the dog. *Circ. Res.* 25:153.

Gray, J. S., 1950. *Pulmonary Ventilation and Its Regulation.* Charles C Thomas, Springfield, Ill.

Guyton, A. C., and Coleman, T. G., 1969. Quantitative analysis of the pathophysiology of hypertension. *Circ. Res.* 24:1, Suppl. 1.

Guyton, A. C., and Crowell, J. W., 1961. Dynamics of the heart in shock. *Fed. Proc. 20:51.*

Guyton, A. C., and Crowell, J. W., 1964. Cardiac deterioration in shock. *Intern. Anesthesiology Clinic—Shock* 2:159.

Holloman, G. H., Jr., Milhorn, H. T., Jr. and Coleman, T. G., 1968. A sampled-data regulator for maintaining a constant alveolar CO_2. *J. Appl. Physiol.* 25:463.

Milhorn, H. T., Jr., 1966. *The Application of Control Theory to Physiological Systems.* W. B. Saunders Company, Philadelphia.

Milhorn, H. T., Jr., 1969. Practical applications of control theory in respiratory and circulatory research. *Fed. Proc.* 24:84.

Sagawa, K., Ross, J., and Guyton, A. C., 1961a. Quantitation of the cerebral ischemic pressure response in dogs. *Am. J. Physiol.* 200:1164.

Sagawa, K., Taylor, A. E., and Guyton, A. C., 1961b. Dynamic performance and stability of the cerebral ischemic pressure response. *Am. J. Physiol.* 201:1164.

Wiener, N., 1961. *Cybernetics.* John Wiley and Sons, New York.

11

Present Concepts and Future Directions of Research on the Control of Metabolism

Arthur L. Black and Michael L. Bruss

Numerous factors are now recognized that play a role in the regulation of metabolic processes in the animal. These factors encompass a broad spectrum ranging from behavioral activities of animals, to complex physiological processes like respiratory and circulatory controls, down to the molecular level where various compounds influence individual enzymes that enhance the release of chemical energy and catalyze biosynthetic activities inside the cell. Any effort to consider metabolic control in one chapter must necessarily be restricted to selected examples and certainly will result in some oversimplification of the highly diverse and complex interactions that are actually involved in animal metabolism.

Some of the most impressive progress in recent years has occurred in the understanding of processes involved in regulating biochemical pathways in the cell. This chapter will consider some of these processes, reviewing current concepts on how the molecular controls are effected. It will identify some of the gaps in our understanding, indicate where progress is needed to enhance our ability to influence metabolic controls, and point out an important contribution that scientists can make in aiding these developments.

It appears that we are at the forefront of some major breakthroughs in understanding biological regulations, and that these developments should be enormously beneficial for mankind. For example, the possibility exists for controlling metabolic processes in a way that will result in more efficient animal production. This does not mean that control processes, as they normally function, are inefficient. Their success for animal survival is undeniable since they have been tested and proven under open competition from environmental challenges during long periods of evolutionary development. However, as Hale pointed out (1974), some of the more difficult challenges have resulted from loss of food supply, threatening starvation, or the presence of predators which jeopardizes individual survival, and various problems of reproduction which threaten survival of the species. When the animal is isolated from these types of "threats," some of the metabolic controls that have developed to counter these challenges would be superfluous to ensure survival of the individual and the species, and consequently some energetically costly metabolic processes might be eliminated, making the animal more efficient. It would then be enabled to grow at a more rapid rate or to produce its food or fiber products more efficiently with respect to its dietary intake.

The level of dietary intake directly affects productive capability. It is obviously important to fully understand the many physiological and environmental factors that influence appetite (Baumgardt, 1974) and to continue development of methods for affecting food intake to maximize performance.

Much remains to be done before it will be possible to "throw the biochemical switches" and selectively divert the metabolic fluxes to achieve more efficient utilization of nutrient material. For one thing, a technique is needed which will accurately measure the flux of metabolites along the various biochemical pathways and provide a basis for assessing the metabolic fate of substances in the intact animal under different conditions. An approach that the scientist can use to monitor and identify conditions for maximum efficiency of nutrient utilization will be presented at the end of this chapter.

Regulation of Enzyme Activity

Enzyme activity is one of the major factors influencing the magnitude of metabolic fluxes in the animal. It may be regulated on a long-term basis by changing the amount of enzyme in the cell or sharply by changing the kinetic properties of specific enzymes. Since enzymes function as catalysts, only small amounts of material must be changed in order to effect large changes in the mass of material that can be processed along a given pathway. For both types of regulatory processes, examples will be given below with references to review articles for more complete information.

Control of Protein Synthesis

The synthesis of apoenzyme involves a large number of steps, many of which may be regulated to affect enzyme activity in the cell. Information is only beginning to develop on how these controls function in animals, so continued research will be necessary before a clear picture of regulation at the gene level emerges.

The first successful model to account for increases (induction) and decreases (repression) in amount of enzyme in cells, proposed by Jacob and Monod (1961), incorporates many of the experimental observations that have been made with microorganisms. Their structural-gene hypothesis states that the information contained in deoxyribonucleic acid (DNA) is necessary and sufficient to specify the structure of all cellular protein. They proposed a hypothetical message, subsequently identified as messenger ribonucleic acid (mRNA), which transfers information from DNA to the cytoplasmic components that assemble the peptide chains. The message is very labile and, thus, its rate of synthesis directly determines the rate of protein synthesis. The DNA is complex, with several components that interact in the formation of mRNA. Regulatory genes affect the formation of a repressor substance, subsequently identified as a protein molecule (Riggs and Bourgeois, 1968), which acts on an operator site to prevent associated structural genes from turning out mRNA. Substances which act as inducers of enzyme activity function to inhibit the action of the repressor, thus freeing the operator gene and its group of structural genes. The "lac" operon of *E. coli* has been extensively studied, and illustrates the inducing action of lactose in "turning on" the synthesis of ß-galactosidase and other enzymes affecting lactose metabolism. There are many examples of repression resulting from the accumulation of an amino acid which functions as a co-repressor (Cohen, 1965). The amino acid and repressor protein inhibit a specific operator gene and "turn off" the structural genes responsible for mRNA that specifies apoenzyme molecules to catalyze the sequence of reactions leading to synthesis of the amino acid. Regulation of protein synthesis at this level is referred to as transcriptional control.

There are many characteristics of enzyme regulation in animal cells that cannot be accounted for by the Jacob-Monod model. In contrast to microorganisms where the mRNA has a short half-life, comprising a few minutes, the half-life of mRNA in animal tissues is measured in hours. Also characteristic of higher animals is that only a small part of the RNA formed in the nucleus ever reaches the cytoplasm, indicating that many RNA molecules are destroyed without functioning as mRNA. Studies on the induction of tyrosine transaminase in rat hepatoma cells treated with adrenal steroids (Tomkins et al., 1969) have uncovered fur-

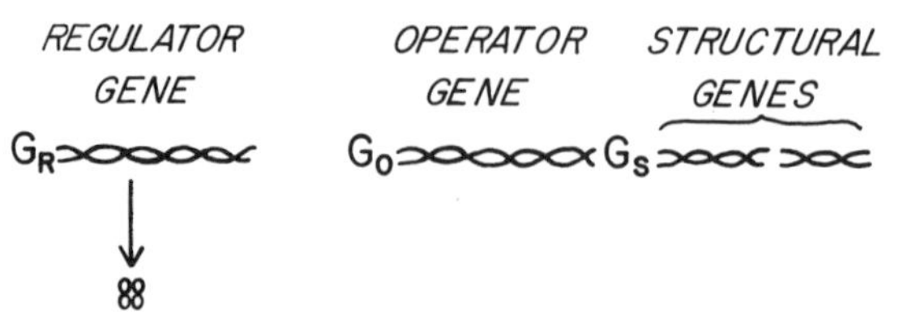

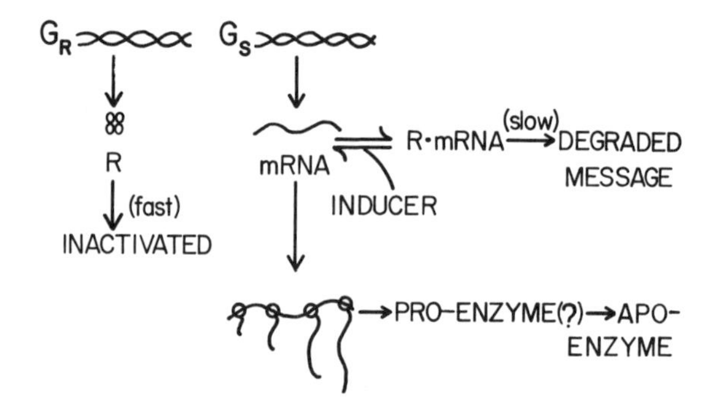

Figure 11-1. Two proposed models to account for regulation of protein synthesis in cells. Above, the Jacob-Monod (1961) model describes transcriptional control of protein synthesis in prokaryotic cells. Below, the model for post-transcriptional control (Tomkins et al., 1969) accounts for several of the characteristic responses of eukaryotic cells as demonstrated in cell culture with a derived line of rat hepatoma cells. The inducer substance in these studies was a glucocorticoid, either dexamethasone or cortisol; tyrosine transaminase was the enzyme induced.

ther characteristics of enzyme regulation in mammalian cells. After addition of cortisol or dexamethasone to hepatoma cells, the enzyme level increased five- to fifteenfold within 10 hours. Following induction, removal of the cortisol by dilution caused a loss of enzyme activity in the cells. Subsequent addition of Actinomycin D (Act D) immediately reversed the drop in tyrosine transaminase activity, but, the effectiveness of the Act D decreased when it was added at later intervals after removal of the inducer.

Tomkins et al. (1969) have proposed a model for post-transcriptional control of mRNA activity which accounts for these various observations (see Figure 11-1). The model includes a regulatory gene controlling the production of a *highly* labile repressor, and a structural gene that constantly turns out mRNA for tyrosine transaminase. In the absence of inducer, the cell is turning out both repressor and mRNA for the enzyme. The repressor, presumably a protein, holds the messenger in an inactive state and unavailable for translation into apoenzyme at the ribosome level. In combination with repressor, the mRNA undergoes slow degradation and is lost from the cell. The sole function of the inducer, cortisol, is to antagonize the post-transcriptional repression by restoring mRNA to the active state, allowing translation to proceed with ensuing formation of active enzyme molecules.

This interesting model provides one possible mechanism for hormonal induction of enzyme synthesis in animal cells. It is obvious, however, that there are many additional possibilities for control. The interaction of mRNA with the ribosome involves a complex of factors for initiating protein synthesis. Additional factors are required during peptide synthesis in the process of chain elongation, and in chain termination when the messenger has been completely translated (Lengyel, 1969). Most of these factors have not been fully identified and little, if anything, is known about their regulation. The opportunity also exists for influencing protein synthesis after translation is completed, in a post-translational modification of the peptide chain. Thus, much remains to be done before we will have the complete picture on the processes responsible for changing the rate of synthesis of new enzyme molecules in the cell, but current studies adumbrate the possibility for future control of metabolic fluxes by intervening ("throwing switches") at the gene level of cell function.

Enzyme Regulation by Protein-Protein Interaction

Patton (1974) has discussed interaction between protein moieties in relation to cell membrane functions. This type of process offers a good possibility for metabolic regulation not only at the membrane level, where permeability influences metabolite access to the cell, but also at the enzyme level. The lactose synthetase enzyme, which is responsible for synthesis of milk sugar in the mammary gland (Watkins and Hassid, 1962), illustrates how this type of regulation might function. Brodbeck et al.

(1967) discovered that the enzyme was a dimer of which one component was the whey protein, α-lactalbumin. The latter has been designated a "specifier protein" (Brew et al., 1968), to reflect its unique role in determining the final enzyme activity. The other moiety of lactose synthetase, called the "A" component, is actually a galactosyl transferase, distributed widely throughout the body where it catalyzes the transfer of uridine diphosphogalactose (UDP-gal) to a variety of complex carbohydrate derivatives. For example, in glycoprotein synthesis, UDP-gal is transferred to an N-acetyl-hexosamine. The "A" protein is involved in formation of a structural type of protein in all tissues, including mammary gland. With the onset of lactation, α-lactalbumin appears and changes the specificity of "A" protein to catalyze only the reaction:

$$\text{UDP-gal} + \text{glucose} \rightarrow \text{lactose} + \text{UDP}.$$

It can be shown that if α-lactalbumin is added to liver tissue, it combines with the "A" enzyme to become lactose synthetase, causing the liver tissue to produce lactose. However, mammary gland is the only tissue that synthesizes α-lactalbumin, and therefore the only place for lactose synthesis in the intact animal. This interesting mechanism illustrates how the cell can develop an entirely new enzyme capability during differentiation by redirecting the activity of an enzyme (galactosyl transferase) to a new function (lactose synthetase) on the appearance of a specific protein (α-lactalbumin). The net effect is to change the metabolic fate of galactose in the cell by diverting it from glycoprotein synthesis to lactose synthesis.

Turkington and Hill (1969) have demonstrated experimentally that progesterone concentration controls α-lactalbumin synthesis. They have proposed that high hormone levels circulating during pregnancy inhibit the synthesis of α-lactalbumin, but as the hormone level decreases at term, the inhibition is removed and mammary cells begin synthesizing lactose.

By itself, α-lactalbumin has no enzymatic activity, and no other function has been described for it besides its acting as a specifier protein to change the catalytic nature of the "A" enzyme. This type of protein-protein interaction might be one of the more important types of metabolic control processes, but additional work will be necessary to determine how one might intervene to influence this type of reaction in a desirable way.

Protein Turnover in the Cell

Schimke (1969) has pointed out that animal as well as bacterial cells respond to the presence of inducers by increasing the level of specific enzymes in the cell. However, in bacterial cells the total amount of a given enzyme remains constant in the absence of inducer, while in animal cells the enzyme level decreases exponentially once the inducer is removed. The enzyme content of animal cells is in a dynamic steady state resulting from an equilibrium between the rate of synthesis and intracellular degradation of enzyme molecules. The final level of enzyme is determined by the ratio of these two processes, each of which can be varied independently. Techniques for immunological isolation of enzyme molecules combined with tracer studies have been used to establish turnover rates for several enzymes in the cell. Studies with tryptophan pyrrolase (Schimke et al., 1965) and agrinase (Schimke, 1962a) have served to identify control mechanisms functioning under different experimental conditions. In the case of tryptophan pyrrolase, the enzyme initiating the degradation of tryptophan, an increase in enzyme content occurred after animals were given either tryptophan or cortisol. When both substances were given together, the response was greater than with either alone, demonstrating that each functioned independently to increase the level of tryptophan pyrrolase. The increase following cortisol treatment was due to an increase in the rate of synthesis of the enzyme, while tryptophan (as substrate) functioned to stabilize the enzyme and reduce its rate of degradation.

Arginase is one of five enzymes involved in the urea cycle, which converts amino-acid nitrogen into urea during catabolism of amino acids. Schimke (1962a) has shown that all five of these enzymes respond in parallel during changes in protein catabolism, but he has concentrated on arginase to elucidate the processes responsible for regulating its level in the cell. Rats fed a high-protein diet (60% casein) had high levels of urea-cycle enzymes, with the level of arginase two to three times higher than normal. When dietary protein was reduced to 8%, the level of arginase and other urea-cycle enzymes decreased. This change was the result of both decreased synthesis and increased destruction of arginase to bring it to a new steady-state level. When rats were starved, protein catabolism increased since the protein served both as a source of energy and for gluconeogenesis. Under these conditions, arginase level was increased by maintaining constant the rate of synthesis and decreasing to zero the rate of degradation of the enzyme. The molecular processes responsible for these adjustments remain unknown. It seems likely that factors affecting the rate of enzyme synthesis would intervene at the level of transcription or translation while factors that affect enzyme destruction presumably function at the level of proteolytic enzymes, perhaps involving the lysosomes of the cell. Future studies will identify the mechanisms, but it is clear that controls operate both on synthesis and degradation to effect changes in the amount of specific enzymes in the cell.

"Acute" Regulation of Enzyme Activity

The mechanisms discussed above provide the bases for relatively slow or "chronic" adjustments of metabolic control. The time for response varies from a few hours in the case of enzymes of short half-life, such as tyrosine transaminase or tryptophan pyrrolase, up to several days for some of the more stable enzymes, like arginase (Berlin and Schimke, 1965).

The cell also has the capacity for making very rapid or "acute" adjustments in enzyme activity by activating or inactivating enzyme molecules. Many enzymes subject to this type of metabolic control have been identified throughout the major anabolic and catabolic pathways of the cell. A few of these control points are indicated in an abbreviated metabolic scheme (Figure 11-2), which illustrates the diversity and multiplicity of biochemical controls. Selected examples will be discussed below to illustrate the major types of control which enable the cell to make rapid adjustments in metabolic fluxes in response to changing needs, and to identify the biochemical nature of the signals that cells recognize. More complete information on biochemical controls can be obtained in several excellent reviews (Newsholme and Gevers, 1967; Stadtman, 1966; Scrutton and Utter, 1968; Villar-Palasi and Larner, 1970; Schimke and Doyle, 1970).

Positive and Negative Feedback Effectors

Feedback inhibition enables the cell to shut off the flow of metabolites along a given pathway when an adequate level of product formed by the pathway has accumulated. Numerous examples are known where this type of regulation occurs in microbial cells for pathways involved in synthesis of amino acids and nucleotides (Cohen, 1965; Stadtman, 1966). In general, as the product accumulates, it inhibits the first step unique to the reaction sequence.

This type of regulation in the animal cell can be illustrated by reactions involved in glucose metabolism. For example, glucose itself (Segal, 1959) as well as inorganic phosphate (Nordlie and Arion, 1965) inhibit the glucose-6-phosphatase (G6P) of liver. Thus glucose release from the cell is reduced when there is already a high level in the plasma. The reverse reac-

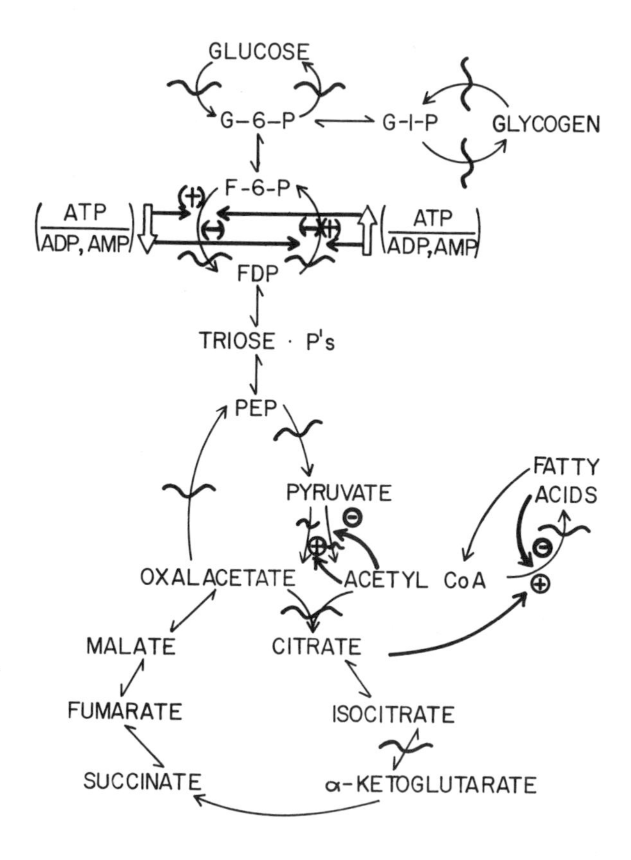

Figure 11-2. Scheme indicating points of metabolic control (~) in pathways involved in carbohydrate and lipid metabolism. Sinusoidal lines designate points of regulation; arrows to the control points show where metabolites exert positive (plus sign) or negative (minus sign) effector action to influence the metabolic flux along the pathway. The ratio (ATP/ADP, AMP), the "adenylate charge" discussed in the text, is only one of several factors influencing the flux of glycolysis and gluconeogenesis in the animal.

tion, catalyzed by hexokinase in peripheral tissues, is inhibited by G6P (Crane and Sols, 1953), reducing the uptake of glucose when the cell has acquired sufficient glucose or other metabolizable substrate that leads to increased levels of G6P. The net effect is to conserve glucose for nerve tissue when it is present in limiting amounts—e.g., in starvation, and the animal is utilizing fatty acid reserves to meet its energy needs.

Another example of feedback inhibition occurs during fatty acid synthesis. Fatty-acyl-CoA, a metabolically active form of long-chain fatty acids, inhibits the acetyl-CoA carboxylase, the first and rate-limiting step of a complex enzymatic sequence involved in converting acetyl-CoA moieties into fatty acids in the cell (Numa et al., 1965). The fatty-acyl-CoA is a precursor for triglyceride synthesis, and is also an end product of a long series of steps involved in fatty acid synthesis. By inhibiting the first step unique to fatty acid synthesis, the cell guards against the accumulation of many metabolic intermediates after adequate product has been formed. The advantage of this arrangement is readily apparent when one considers the problem that would result if metabolic products inhibited a late step in a serial-reaction sequence. The large pools of intermediates which could accumulate would not only be uneconomical but also pose additional problems of disposal or perhaps even loss of metabolic energy from the cell.

In addition to feedback inhibition, which shuts off a pathway once an end product has accumulated to a certain set level, there is also feed-forward activation. This action can be illustrated by citrate, a positive effector of acetyl-CoA carboxylase, which activates the enzyme and increases fatty acid synthesis (Vagelos et al., 1963).

Citrate and fatty-acyl-CoA are examples of allosteric effectors. These substances interact with an enzyme at some site other than the catalytic site, where the substrate is bound, and affect its kinetic properties. Allosteric effectors may change either the maximum velocity (V_{max}) of an enzyme, its Michaelis constant (K_m), or both factors. Increasing the V_{max} or decreasing the K_m makes the enzyme more effective at substrate concentrations below saturating levels. These responses can occur rapidly, resulting in acute changes in metabolic fluxes. When there is an increase in the enzyme V_{max}, it is as if the cell had increased its total enzyme content although, in fact, no change in rate of synthesis or degradation of enzyme is involved in this type of response.

Control in Response to Energy Change in the Cell

Metabolic processes are especially sensitive to levels of ATP, the molecule which effectively functions as the "energy currency" of the cell. The reason is that ATP is widely active in almost all types of metabolic reactions and provides an immediately accessible reserve of chemical energy (Lehninger, 1971).

The "adenylate charge" of the cell (Atkinson, 1968), reflecting the concentration of ATP in relation to that of ADP and AMP, regulates the direction of metabolic flux at several biochemical steps. Its action in general can be illustrated by the effect on phosphofructokinase and fructose diphosphatase (FDPase), as shown in Figure 11-2. Changes in the adenylate charge reciprocally affect these two enzymes which catalyze the interconversion of fructose-6-phosphate (F6P) and fructose-diphosphate (FDP) by an allosteric mechanism. When the charge is high, there is an inhibition of phosphofructokinase together with an enhanced activity of FDPase, a change that favors gluconeogenesis and depresses glycolysis. This action can also be viewed as feedback inhibition since one function of glycolysis is to provide ATP for the cell, and as its level increases, the need for additional amounts decreases, so the glycolytic flux, which produces ATP as one product, is throttled down. When the adenylate charge in the cell falls, the opposite changes occur, leading to an inhibition of the FDPase and enhancement of the phosphofructokinase. Then greater production of ATP by glycolysis (and also the tricarboxylic acid cycle) is favored until the adenylate charge is restored. At the same time, the inadequate ATP level depresses biosynthetic reactions, as illustrated in this example by the decreased activity of FDPase, reducing the formation of G6P, the precursor of glucose or of glycogen. The importance of this reciprocal action of metabolic signals on enzymes catalyzing reversible steps around a single reaction has been emphasized by several investigators (Krebs et al., 1964; Atkinson, 1966; Scrutton and Utter, 1968). In effect, it avoids a fruitless short circuit which would waste ATP by repeatedly interconverting F6P and FDP with a net loss of one molecule of ATP per conversion. This type of reciprocal action is at the heart of metabolic control at several points in the metabolic scheme, and accounts for the very strong directive influence of effectors which change the metabolic flux along pathways under different conditions. The Teppermans (1970) have characterized this coordinated control as "the Sherringtonian metaphor."

Pyruvate carboxylase, a key enzyme in gluconeogenesis which is allosterically affected by acetyl-CoA levels (Utter and Scrutton, 1969), link carbohydrate and lipid metabolism in the cell. This enzyme catalyzes the conversion of pyruvate into oxalacetate which initiates, probably with an important rate-limiting reaction for, gluconeogenesis from pyruvate, lactate, and some of the amino acids. Pyruvate carboxylase has an absolute requirement for acetyl-CoA, and the K_a is close to the measured concentration at which acetyl-CoA is found in the cell, making the enzyme especially sensitive to fluctuations in acetyl-CoA level. At the same time, acetyl-CoA is an allosteric inhibitor of pyruvic dehydrogenase (see Figure 11-2). The effect of this reciprocal interaction is to reduce the conversion of pyruvate into acetyl-CoA, the normal pathway which it follows when carbohydrate is being utilized for chemical energy, whenever acetyl-CoA levels increase in the cell. This situation arises under various conditions—e.g., starvation, diabetes, or whenever inadequate carbohydrate reaches the animal's cells, and fat reserves are utilized to meet metabolic needs. Beta oxidation of fatty acids, which produces acetyl-CoA as the end product, enhances the flux of pyruvate toward pyruvate carboxylase and gluconeogenesis. Thus glucogenic precursors are conserved at a time when their supply is limited and when glucose concentration could otherwise fall to critical and physiologically harmful levels. It is readily apparent how this type of interaction would be beneficial to the animal under conditions of starvation or in other situations where glucose would be a limiting factor—e.g., confinement to a high-fat diet. Such metabolic situations also increase the activity of phosphoenolpyruvate-carboxykinase, FDPase, and G6Pase, all of which favor the flow of metabolites toward gluconeogenesis to avoid hypoglycemia.

In-Vivo Evidence of Metabolic Control in Ruminants

It is possible to demonstrate control processes in the intact animal and show that they do respond to changes in circulating levels of normal metabolites, as one would predict from established molecular mechanisms. One example is the effect of increasing plasma butyrate concentration on gluconeogenesis from pyruvate-C^{14} and also from C^{14}-labeled carbon dioxide in lactating dairy cows. Butyrate is one of the short-chain fatty acids produced by the rumen microorganisms as a normal end product of their fermentation processes. It is partially converted into ketone bodies during absorption (Pennington, 1952) by rumen epithelium, but the remainder is taken up by the liver where it undergoes ß-oxidation and contributes to the acetyl-CoA pool. The avidity of liver uptake is demonstrated by the fact that little of the butyrate passes the liver and appears in the peripheral circulation (Annison et al., 1957).

Butyrate was used experimentally to increase the acetyl-CoA pool in the liver by injecting 0.3 to 1.0 mmole Na butyrate per kilogram body weight into the blood of cows over a 10-min interval. This treatment resulted in an increased transfer of C^{14} from pyruvate-2-C^{14} into plasma glucose and diverted the flux of pyruvate carbon from acetyl-CoA (pyruvate dehydrogenase activity decreased) toward oxalacetate (pyruvate carboxylase activity increased) (Black et al., 1966). The latter effect was shown by the change in the intramolecular C^{14}-distribution in glutamate after butyrate injection. Glutamate is derived from a tricarboxylic acid cycle intermediate, α-ketoglutarate, and can be used as an index of the relative amount of C^{14} from pyruvate-2-C^{14} entering the cycle as acetyl-CoA versus that entering as oxalacetate. In two normally fed cows, 50% of the pyruvate carbon had entered the cycle as oxalacetate, but when the cows were injected with butyrate the flux via oxalacetate increased so that 80 to 90% of the total pyruvate carbon that had entered the cycle had been introduced via oxalacetate. Butyrate caused the flux via oxalacetate to increase by 40 to 75% over that observed when the cows received no butyrate.

A third cow (Cow No. 1050; see Black et al., 1966) had been fasted for 24 hours before measurements were made on her pyruvate metabolism. In the absence of butyrate, her flux via oxalacetate was already higher than in the fed cows, presumably reflecting increased nonesterified fatty acid (NEFA) levels, expected in fasting animals. On reaching the liver these NEFAs could produce an increase in acetyl-CoA levels and stimulate gluconeogenesis. Prior to butyrate treatment, 80% of the pyruvate flux had entered the cycle via oxalacetate, but after injection of butyrate, the flux via oxalacetate increased to 95% of the total pyruvate carbon entering the cycle.

The effect of butyrate on gluconeogenesis was substantiated by measuring the transfer of C^{14} into plasma glucose during constant infusion of cows with $C^{14}O_2$ (Anand and Black, 1970). The specific activity of C^{14} in the circulating glucose in relation to that in $C^{14}O_2$, called the transfer quotient (Kleiber and Black, 1956), provides an index of the rate of gluconeogenesis in the intact animal. The measured transfer quotient, after isotope equilibrium had been established, showed that 10% of the glucose carbon was coming from CO_2 Since the maximum amount of glucose carbon that can be derived from CO_2 is theoretically 16.7%, the observed results indicate that the cows were deriving about 60% of their glucose flux ($10/16.7 \times 100$) from gluconeogenic reactions involving CO_2 fixation. When the cow was given a pulse load of propionate (0.7 mmole/kg over 10 min), there was a rapid increase of plasma glucose concentration and a 40 to 60% increase in the transfer quotient. The latter can be explained as a mass-action effect of increased substrate on propionate metabolism, but it might also reflect an activation of liver pyruvate carboxylase by increased levels of propionyl-CoA (Utter and Scrutton, 1969).

When the cow was pulse-loaded with an equal amount of butyrate, the same response (increased glucose concentration and increased transfer quotient) was elicited. It has been clearly established that butyrate is not a glucogenic precursor in the cow (Black et al., 1961), nor in sheep (Leng and Annison, 1962, 1963), so its action in increasing gluconeogenesis must be indirect. The most reasonable explanation for butyrate's action is that discussed above, that butyrate "turns on" the pyruvate carboxylase enzyme by increasing the concentration of the positive effector substance, acetyl-CoA, in the liver. Even if this explanation proves to be an oversimplification of the butyrate effect, the results do provide experimental evidence that metabolic processes can be influenced in a predictable fashion by changing levels of normal metabolites in the intact animal. In this case, the gluconeogenic flux was increased within minutes after butyrate was given, and was maintained at an elevated level for an interval of 40 min. These experimental results support the major thesis of this chapter, that metabolic fluxes in the animal can be influenced by normal metabolites. As we learn more about them and understand their control, we will be in a position to influence metabolic processes in a desirable manner. This capability should help to produce animal products more efficiently.

Enzyme Activation by Covalent Changes

There is one other important type of metabolic control that should be mentioned because it involves a different mechanism from the allosteric interactions already described. Several enzymes undergo change in activity as a result of covalent modification of the apoenzyme molecule. This type of control can be illustrated by the enzymes affecting turnover of glycogen stores in the body. The reactions involved are extremely complex since controls appear at several different levels of interaction, and the reactions in the liver differ from those occurring in muscle. The reactions in liver appear to be somewhat less complicated.

Glycogen breakdown is controlled by phosphorylase which can exist in either an active form, phosphorylase *a,* or in an inactive form, phosphorylase *b*. Glycogen synthesis is controlled by a different enzyme, glycogen synthetase, which also exists in two forms, one of which, synthetase *a* (also called I-form), is most active. The direction of the flux between glycogen and G6P will depend on the activity of phosphorylase in relation to glycogen synthetase. The amount of each enzyme in the active form is determined in turn by the activity of two kinase enzymes and two phosphatase enzymes which interconvert the active and inactive forms of phosphorylase and glycogen synthetase (see Figure 11-3). The kinase enzymes utilize ATP to phosphorylate a serine residue on phosphorylase *b* (inactive form) and glycogen synthetase *a* (active form). The effect of this modification is to turn-on phosphorylase and to inactivate glycogen synthetase. The two phosphatases remove the activating phosphate group from the serine residues, and, thereby, inactivate phosphorylase and activate glycogen synthetase. A variety of effectors impinge on these kinases and phosphatase enzymes, and indirectly cause either glycogen synthesis or breakdown.

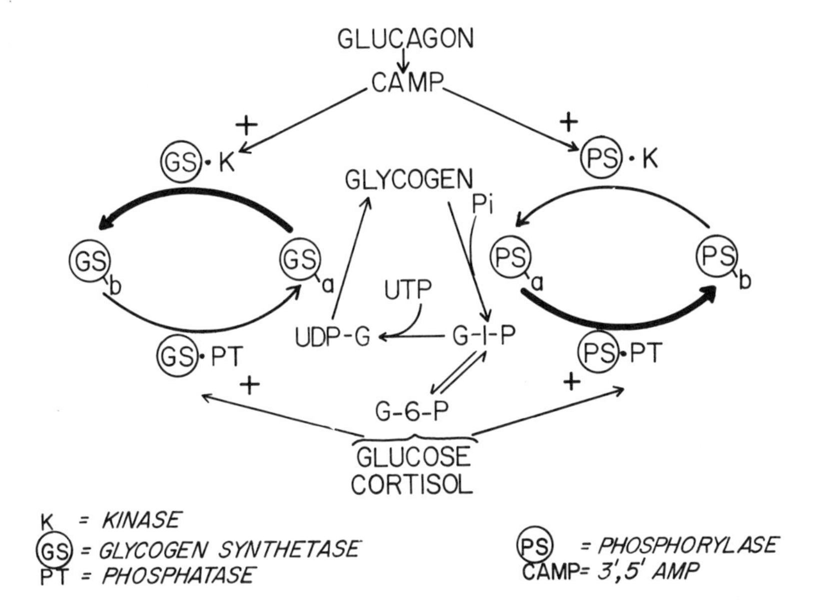

Figure 11-3. Abbreviated scheme to illustrate the complex nature of regulation involved in liver glycogen metabolism. Several factors, in addition to glucagon, affect the level of cAMP in liver cells and influence the rate of glycogen turnover.

Some of the recognized effectors, shown in Figure 11-3, illustrate the very complex nature of the regulation of glycogen metabolism. Glucagon is one of the hormones that causes rapid breakdown of liver glycogen stores. It functions by activating an enzyme, adenyl cyclase, in the liver cell membrane, which converts ATP into 3′,5′-AMP (cyclic AMP, or cAMP). cAMP activates both kinases activating phosphorylase and inactivating glycogen synthetase. The net result is the release of glucose into the blood and the depletion of liver glycogen stores. There are additional factors that affect the cell concentration of cAMP (Robinson et al., 1968) and play a role in the rate and extent of glycogenolysis.

In the presence of high cortisol levels, the situation can be reversed: enhanced activity of a phosphatase which activates glycogen synthetase and inactivates phosphorylase accounts, in part, for one of the early effects of glucocorticoids which leads to increased liver glycogen levels.

It should be emphasized that in the regulatory processes affecting glycogen metabolism, the total amount of phosphorylase or glycogen synthetase does not change during these acute responses, only the relative amount of each enzyme in the active and inactive form. The control of glycogen turnover also illustrates a general principle in metabolic regulation, that when there are reversible reactions between metabolites that involve changes in energy levels (e.g., that utilize high-energy phosphate bond potential in one direction), the mechanisms are always coupled so that one direction is inhibited when the other direction is activated.

More complete discussion of glycogen metabolism and its regulation can be obtained from recent reviews (Villar-Palasi and Larner, 1968, 1970).

V_{max} as an Index of Changes in Metabolic Flux

All techniques have their limitations, and it is important for investigators to be aware of them in order properly to assess the significance of experimental results. It is especially important to consider the latent possibility of artifacts in analytical procedures that can mislead the investigator and vitiate hypotheses. There is good reason to believe that such has not always been done in assembling the evidence on control mechanisms and their metabolic effects. Much of our current picture on regulation is based on measured changes in tissue enzymes assayed under zero-order kinetics to establish their V_{max} Several investigators have expressed concern about the possibility that assay conditions might be so different from conditions in the animal that they do not reflect meaningful measurements of *in vivo* capabilities. Thus, the question arises whether one can use changes in enzyme levels as direct and unambiguous evidence of changes in the metabolic flux along the pathway involved.

Scrutton and Utter (1968) have compiled several examples to illustrate the point that measured value for V_{max} cannot be applied directly to assess the metabolic flux *in vivo*. For example, the V_{max} for two of the glycolytic enzymes, phosphofructokinase (PFK) and pyruvic kinase (PK), were 10 to 100 times greater than the measured rate of glucose utilization in each of several different rat tissues. Thus, one would overestimate glucose metabolism significantly if he based his evaluation solely on the activities of selected enzymes. An explanation for these discrepancies between V_{max} and metabolic flux is apparent if one recognizes the possibility for overall control of net rates of flux at other metabolic points—e.g., membrane transport of glucose or, at the initial activation, glucokinase (liver) or hexokinase (muscle and other peripheral tissues). Any of these points may limit the glucose flux through the glycolytic pathway to levels below the maximum capability of other controllable enzymes along the path. The discrepancy might also reflect a reserve capacity of PFK to utilize hexose moieties derived from glycogenolysis under stressful or emergency conditions. However, the V_{max} for PFK exceeded by nearly fivefold the μmmole of glucose utilized by glycolysis in the tetanized muscle.

Ashida and Harper (1961) have demonstrated a positive correlation between arginase levels in rat liver and greater urea excretion as dietary protein increased. Thus, there appears to be a direct link between the amount of protein being degraded and the capacity to convert amino nitrogen into urea. Schimke (1962a) has further shown that all five of the urea-cycle enzymes undergo coordinated change, increasing in parallel with greater protein intake. This synchronized adaptation of the urea-cycle enzymes is not seen invariably, for when animals were starved arginase remained constant (Schimke, 1962b) or even decreased (Freedland and Sodikoff, 1962); when the animals were fed an arginine-deficient diet (Schimke, 1963), arginase decreased while the remaining enzymes increased, presumably to provide the animal with a greater capacity to synthesize arginine. These changes all make good sense, and it is easily understandable how these capacities might confer survival advantage on the animal faced with the unpredictable nature of environmental factors.

One aspect remains especially puzzling, however. The calculated capability for urea formation, based on arginase assay in the liver of rats fed diets with increasing protein content, indicated that the rat fed a diet containing 45% protein was theoretically able to produce an amount of urea each day equal to its body weight. Furthermore, when dietary protein was increased to 70%, the arginase increased an additional 30%. It seems obvious that a capability for urea synthesis like that suggested by the arginase level at the 45% protein diet, would never be used, so why should the enzyme level continue to rise with greater protein intake? There are various possibilities that might account for this situation. It is likely that many, if not all, enzymes do not function in the cell at zero-order kinetics, as measured in enzyme assays. Thus, substrate is often limiting the rate of the reaction in the cell. It is also possible that the enzyme may be altered during extraction so that the kinetic parameters measured *in vitro* do not apply *in vivo*.

The difficult problem of interpreting the significance of V_{max} unambigously can be illustrated by a series of investigations in Lardy's labocaratory on factors affecting the activity of phosphoenolpyruvate-carboxykinase (PEPCK). When L-tryptophan was administered to rats, the activity of the gluconeogenic enzyme PEPCK was markedly enhanced (Foster et al., 1966). However, tryptophan had no effect on PEPCK when added *in vitro,* and, paradoxically, *in vivo* tryptophan inhibited the glucogenic effect of cortisol and prevented glycogenesis from pyruvate, malate, and aspartate, but not from glycerol or glucose (Ray et al., 1966). Thus, on the one hand, tryptophan seemed to turn on gluconeogenesis (increased V_{max} for PEPCK) yet, on the other, it clearly inhibited gluconeogenesis (decreased utilization of pyruvate, malate, etc.). The latter effect was confirmed by using metabolite cross-over anaylsis which showed that tryptophan administration blocked gluconeogenesis at the level of PEPCK (Veneziale et al., 1967). This paradox was finally resolved when it was shown that the PEPCK assay was artifactual. One of the tryptophan metabolites, quinolinic acid, complexes with a metal ion and PEPCK, inhibiting the enzyme *in vivo.* However, as the tissue is prepared for assay, the inhibitor is diluted off, leaving the enzyme in an activated state with a V_{max} twice as great as that of the nonactivated enzyme.

Another example of some dichotomy between measured V_{max} and metabolic flux can be drawn from enzyme data obtained by Baldwin (1966) in mammary gland tissue. It has been shown that enzyme levels in the mammary gland of the rat and the guinea pig (Abraham and Chaikoff, 1959; McLean, 1958; Baldwin and Milligan, 1966) increase markedly, beginning with the initiation of lactation. Baldwin's study (1966) confirmed the increase in enzyme levels in the rat mammary gland, but there was no comparable change in the mammary gland of cows during initiation of lactation. While some enzymes increased in activity when the cow started lactating, several were as high prior to parturition as they were during the first several weeks of lactation. He concluded that enzymes may be present but not "expressed" during pregnancy, and his results demonstrate that V_{max} as assayed under *in-vitro* conditions, may bear little relationship to the actual flux of metabolities along the pathway involved.

Srere (1969) has compiled data from several laboratories on the level of enzymes involved in fatty acid synthesis in *E. coli,* and found, surprisingly, that six of the seven enzymes assayed had rates lower than the measured rate of fatty acid synthesis. The enzyme with the lowest V_{max} had a rate nearly a thousand times lower than the rate of fatty acid synthesis. This relationship is more difficult to interpret, and presumably results from experimental difficulties, perhaps due to loss of enzyme activity during extraction from the cell and assay.

Additional difficulties in obtaining V_{max} and interpreting results have been discussed by Atkinson (1966). There are problems of species differences (Scrutton and Utter, 1968), and even marked variations in enzyme levels among strains of the same species (Eggleston and Krebs, 1969). In addition, there are diurnal variations in metabolic processes which must be carefully considered to avoid additional difficulties in relating V_{max} to control processes (Baril and Potter, 1968; Potter et al., 1968; Wurtman, 1969).

Evidence has been cited which supports the view that V_{max} and flux may vary in parallel, and that enzyme assays can provide an index of metabolic fluxes in the animal. However, there are several observations which dispute this correlation and leave doubt concerning the validity of equating V_{max} and flux, under many conditions. Thus it becomes critically important to have some independent methods for assessing metabolic changes, to minimize the possibility of error when interpreting enzyme assays. Among the future needs, one of the most important is the development of techniques that will provide an accurate basis for assessing the flux of metabolites in the intact animal. Until theories developed from *in-vitro* studies can be tested in the normal intact animal, their physiological significance will remain in doubt.

A Technique for Measuring Metabolite Flux in the Intact Animal

The method described below provides one basis for assessing the flux of metabolites in the intact animal and for determining the contribution that specific substances make to the animal's metabolic processes. In principle, the technique may be used for many of the major metabolites in the body, but this illustration uses experimental data obtained after intravenous injection of L-phenylalanine-U-C^{14}. This compound is one of the essential amino acids, and information on its turnover and fate in the animal is of considerable interest. The problem of making these assessments is especially difficult for amino acids because they leave the plasma of the animal so rapidly (Black, 1968), and because of intracellular dilution due to protein turnover. The latter process makes it impossible to use the specific activity of a plasma amino acid to calculate its contribution to the animal's CO_2 production or glucose synthesis.

The technique, a modification of one proposed by Shipley et al. (1967), requires that there be some "sink" in the animal which constantly collects, in an irreversible flow, part of the metabolite flux. The lactating cow provides a suitable system since the mammary gland is constantly drawing off circulating metabolites that are used in synthesis of carbohydrate (lactose), fat, and protein which, in turn, are secreted in milk. A measured amount of C^{14}-phenylalanine was injected into the jugular vein over an interval of 1-2 min to ensure its distribution throughout the circulating plasma pool at the start of the experiment. Subsequently, carbon-14 was measured in respired CO_2, plasma glucose, and the various components of milk. Figure 11-4 shows the system in diagramatic form and illustrates how the milk casein functions as a sink. Since there are 28 gm casein per kilogram of milk, and 100 gm casein contains 5.8 g phenylalanine, each kilogram of milk acts as a sink for 1.62 gm phenylalanine. Some exit arrows represent the flow of carbon from the phenylalanine pool for casein synthesis and other metabolic processes, such as the utilization of phenylalanine's carbon for gluconeogenesis or lipogenesis, or its oxidation to carbon dioxide.

Table 11-1. Recovery of Carbon-14 in Milk Casein after Intravenous Injection of Cows with L-U-C^{14}-Phenylalanine

Experiment number	*Cow number*	*Milk production (kg/day)*	*C^{14} injected (μci)*	*C^{14} in casein* (%)*
1	702	13.4	965	26.1
2	2175	12.4	970	23.8

*Cumulative recovery during seven days, expressed as percentage of C^{14} injected.

Table 11.2. Percentage of Carbon-14 in Amino Acids of Casein

Amino acid	*Specific activity (μci/g atom C)*	*C^{14} (μci/100 g casein)*
Phenylalanine	211.0	66.7
Tyrosine	29.3	9.3
Glutamate	1.0	0.8
Aspartate	0.4	0.08
Serine	0.8	0.14
Alanine	0.5	0.06
Glycine	0.7	0.04
Total		77.1

$$\frac{67\ \mu\text{ci}}{77\ \mu\text{ci}} = 87\%$$ of total casein-C^{14} present in phenylalanine.

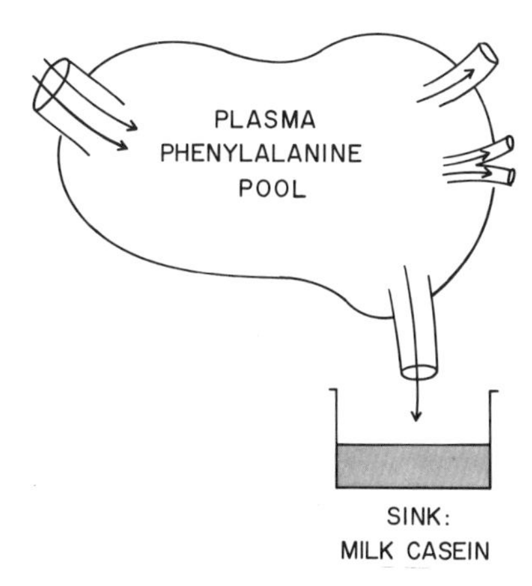

Figure 11-4. Model representing the plasma phenylalanine pool in the cow and the role of milk casein as a "sink" in this system. Arrows entering the pool represent the introduction of new phenylalanine molecules to the plasma from the gastrointestinal tract. Exit arrows represent phenylalanine leaving the pool for other physiological processes.

It was necessary to collect milk until all of the carbon-14 had left the pool, or, by our criterion, until the C^{14} level in casein was not significantly different from background. This required that we collect all milk for seven days and measure the cumulative carbon-14 in the casein formed during that interval. The results obtained with two cows, shown in Table 11-1, indicate that about 25% of the phenylalanine* flux was utilized for casein synthesis. This measurement indicates the efficiency with which the cows utilized the flux through their phenylalanine pools for casein synthesis, and is one metabolic parameter of interest. This value should correspond closely to the efficiency of utilization of "metabolizable" phenylalanine under the conditions of the experiment.

*Since carbon-14 was being followed in this study, the parameter measured was actually the carbon flux through the phenylalanine pool. The nitrogen, hydrogen, and oxygen, which are part of the phenylalanine molecule, are not included in the final assessment. In synthesis of casein, phenylalanine and carbon from phenylalanine are essentially synonymous (but see Table 11-2) and these terms are also synonymous for CO_2 production. However, for synthesis of glucose, both carbon and hydrogen (reducing power) are required, so the C^{14} evaluation measures only part of the contribution that phenylalanine may make to the overall process.

To calculate the flux through the phenylalanine pool, it was necessary to establish how much of the C^{14} in casein was present in amino acids other than phenylalanine. Individual amino acids from casein hydrolysate were separated on ion-exchange columns and their specific activities were determined. The results (Table 11-2), show that most of the C^{14} was in phenylalanine, but that tyrosine had a considerable level and the remaining amino acids only very low amounts. These data demonstrate that about 15% of the tyrosine had been synthesized from phenylalanine in the cow (29.3/211 × 100), and that 87% of the total casein carbon-14 was in phenylalanine. The latter figure is required to calculate the magnitude of the phenylalanine flux as illustrated, using the data from Experiment 1:

(1) 87% of the C^{14} recovered in casein was in phenylalanine.

(2) Casein contained 26.1% of the total C^{14} injected.

(3) Thus (87% × 26.1%) 22.6% of the phenylalanine flux was utilized to provide the phenylalanine moiety of casein.

(4) The daily milk (13.4 kg) contained 21.8 gm of phenylalanine in casein.

(5) The average daily flux through the phenylalanine pool was, therefore:

$$\frac{21.8 \text{ gm}}{22.6} \times 100, \text{ equal to 96 gm phenylalanine per day.}$$

When the same calculation was carried out for Experiment 2, the result was 97 gm phenylalanine per day.

Once the magnitude of the flux has been established it is then possible to calculate the contribution which it makes to other physiological processes. For this assessment, data on the CO_2 flux, are used, but the same approach could be used to calculate the contribution of phenylalanine carbon to gluconeogenesis, lipogenesis, etc.

We have collected total CO_2 for only 3 hours after administering the C^{14}-phenylalanine, so we do not have complete data from our present experiments on the total amount of C^{14} appearing in the CO_2-flux. Assuming that cumulative recovery of C^{14} in CO_2 approached a plateau exponentially, it is possible to estimate the total amount that would have been recovered if the cows had been held in a respiration apparatus for some 7 days. The measured recovery during 3 hours was 7.7% of the total injected, and the extrapolated value was 14.3%. As this estimated value is probably on the low side of the actual value the calculated level of phenylalanine oxidation would be a minimum value. The present data are used to illustrate the technique, with the understanding that complete CO_2 collection would permit a more accurate determination.

The daily fraction of the phenylalanine flux undergoing oxidation (14.3%) amounts to (14.3% of 96 gm/day) 13.7 gm phenylalanine per day. This quantity of phenylalanine would produce 1.09 mole CO_2 per day. The cow was producing CO_2 at the rate of 120 mole/day, so oxidation of phenylalanine comprised (1.09/120) 0.91% of the total CO_2 flux. The same calculation with data from Experiment 2 gave a value of 0.44% of the CO_2 derived from phenylalanine oxidation.

The use of this technique involves the assumption that the cow is in a dynamic steady state with respect to her metabolic turnover. This is probably a reasonable assumption in the adult, nongrowing ruminant, even though it might not be strictly correct. The average pool size should remain reasonably constant in the nongrowing animal, and the amino acid flux through the pool may vary less than it does in monogastric animals (Baril and Potter, 1968). One effect of the rumen is to delay and average out the rate of passage of ingesta through the rest of the gastrointestinal tract, and, in this sense, it buffers the fluctuations in absorption of amino acids from the gut which would tend to maintain a more constant flow through the plasma pool throughout the day.

If later studies show that the assumption of steady-state conditions is not justified, it would be necessary to modify the technique. In that case, administration of the C^{14}-amino acid over a prolonged period, for example, 24 hours, would help to overcome diurnal variations and would give a better estimate of the average daily flux through the pool. The principle, however, remains the same, and could be applied to measure the flux and assess its metabolic fate.

With a technique for measuring metabolic fluxes, one is in a position to test the efficacy of various nutritional regimes, environmental modifications, etc., as reflected in the efficiency of the animal in utilizing some of its major metabolites for production processes. In the example above, it would be possible to modify variables until the utilization of the phenylalanine flux had been maximized for milk protein synthesis and minimized for oxidation, lipogenesis, gluconeogenesis, etc. Ideally, one would hope to identify those conditions that ensure optimum utilization of an amino acid for only those metabolic processes in which it is uniquely required. That would be mainly protein synthesis, but for certain amino acids like glycine and aspartate, it would include nucleic acid synthesis or synthesis of hormones and other essential compounds. At the same time, the ideal conditions would obviate use of the amino acids for body processes that could be provided by other substances less likely to be limiting in the animal. It is wasteful and, therefore, undesirable to oxidize phenylalanine or convert it into lipid since both of these metabolic processes can utilize other substrates.

As we learn more about metabolic controls and how to switch the metabolic fluxes from one fate to another, we should be able to produce animal products even more efficiently than our best efforts now achieve. A great deal remains to be done before this goal can be realized but it is an exciting possibility and worth the effort required to accomplish it.

Summary and Conclusions

In the processes that play a role in controlling metabolic activity in the animal, regulation may occur at several levels, starting at the gene and proceeding through various stages of protein synthesis and turnover. More rapid changes in metabolic activity occur through activation or inhibition of key enzymes along the major metabolic routes by effector molecules which reflect the adenylate charge, the oxidation/reduction potential in the cell, or the accumulation of key metabolites. Some examples illustrate current concepts on the mechanism of action of hormones and metabolic intermediates leading to regulation of metabolic processes.

A technique is described which should be useful for elucidating control processes in the intact animal. It provides a basis for measuring the magnitude of metabolic fluxes and for assessing the metabolic fate of substances in the animal under various conditions.

Through future advances and better understanding of metabolic regulation, it should be possible to influence the metabolic fate of substances in the animal to achieve more effective utilization of essential nutrients and make animals more efficient in the formation of their products.

References

Abraham, S., and Chaikoff, I. L., 1959. Glycolytic pathways and lipogenesis in mammary glands of lactating and non-lactating normal rats. *J. Biol. Chem.* 234:2246.

Anand, R. S., and Black, A. L., 1970. Species differences in the glucogenic behavior of butyrate in lactating ruminants. *Comp. Biochem. and Physiol.* 33:129.

Annison, E. F., Hill, K. J., and Lewis, D., 1957. Studies on the portal blood of sheep. 2. Absorption of volatile fatty acids from the rumen of the sheep. *Biochem. J.* 66:592.

Ashida, K., and Harper, A. E., 1961. Metabolic adaptation in higher animals. VI. Liver arginase activity during adaptation to high protein diet. *Proc. Soc. Exp. Biol. Med.* 107:151.

Atkinson, D., 1966. Regulation of enzyme activity. *Ann. Rev. Biochem.* 35:85.

Atkinson, D. E., 1968. The energy charge of the adenylate pool as a regulatory parameter. Interaction with feedback modifiers. *Biochem.* 7:4030.

Baldwin, R. L., 1966. Enzymatic activities in mammary glands of several species. *J. Dairy Sci.* 49:1533.

Baldwin, R. L., and Milligan, L. P., 1966. Enzymatic changes associated with the initiation and maintenance of lactation in the rat. *J. Biol. Chem.* 241:2058.

Baril, E. F., and Potter, V. R., 1968. Cyclic variation in amino acid transport. *J. Nutr.* 95:228.

Baumgardt, B. R., 1974. Food intake, energy balance, and homeostasis. In *The Control of Metabolism* (J. D. Sink, ed.). The Pennsylvania State University Press, University Park, p. 89.

Berlin, C. M., and Schimke, R. T., 1965. Effect of half-time on time course changes in enzyme levels. *Mol. Pharmacol.* 1:149.

Black, A. L., 1968. Modern techniques for studying the metabolism and utilization of nitrogenous compounds, especially amino acids. In *Isotope Studies of the Nitrogen Chain.* Vienna, p. 287.

Black, A. L., Kleiber, M., and Brown, A. M., 1961. Butyrate metabolism in the lactating cow. *J. Biol. Chem.* 236:2399.

Black, A. L., Luick, J., Moller, F., and Anand, R. S., 1966. Pyruvate and propionate metabolism in lactating cows: Effect of butyrate on pyruvate metabolism. *J. Biol. Chem.* 241:5233.

Brew, K., Vanaman, T. C., and Hill, R. L., 1968. The role of α-lactalbumin and the A protein in lactose synthetase: A unique mechanism for the control of a biological reaction. *Proc. Nat. Acad. Sci.* 59:491.

Brodbeck, U., Denton, W. L., Tanahashi, N., and Ebner, K. E., 1967. The isolation and identification of the B protein of lactose synthetase as α-lactalbumin. *J. Biol. Chem.* 242:1391.

Cohen, G. N., 1965. Regulation of enzyme activity in microorganisms. *Ann. Rev. Microbiol.* 19:105.

Crane, R. K., and Sols, A., 1953. The association of hexokinase with particulate fractions of brain and other tissue homogenates. *J. Biol. Chem.* 203:273.

Eggleston, L. V., and Krebs, H. A., 1969. Strain differences in the activities of rat liver enzymes. *Biochem J.* 114:877.

Foster, D. D., Ray, P. D., and Lardy, H. A., 1966. A paradoxical *in vivo* effect of L-tryptophan on the phosphoenolpyruvate carboxykinase of rat liver. *Biochem.* 5:563.

Freedland, R. A., and Sodikoff, C. H., 1962. Effects of diet and hormones on two urea cycle enzymes. *Proc. Soc. Exptl. Biol. Med.* 109:394.

Hale, E. B., 1974. Beyond the organism—social structure and energy metabolism. In *The Control of Metabolism* (J. D. Sink, ed.). The Pennsylvania State University Press, University Park, p. 195.

Jacob, F., and Monod, J., 1961. Genetic regulatory mechanisms in the synthesis of proteins. *J. Mol. Biol.* 3:318.

Kleiber, M., and Black, A. L., 1956. Tracer studies on milk formation in the intact dairy cow. *A Conference on Radioactive Isotopes in Agriculture*. U. S. Atomic Energy Commission Report No. TID-7512, p. 395.

Krebs, H. A., Newsholme, E. A., Speak, R., Gascoyne, T., and Lund, P., 1964. Some factors regulating the rate of gluconeogenesis in animal tissues. *Adv. Enzyme Reg.* 2:71.

Lehninger, A. L., 1971. *Bioenergetics,* ed. 2. W. A. Benjamin, Inc., New York.

Leng, R. A., and Annison, E. F., 1962. Possible glucogenicity of butyrate in sheep. *Nature* 196:674.

Leng, R. A., and Annison, E. F., 1963. Metabolism of acetate, propionate, and butyrate by sheep liver slices. *Biochem. J.* 86:319.

Lengyel, P., 1969. The process of translation as seen in 1969. *Cold Spring Harbor Symp. Quant. Biol.* 34:828.

McLean, P., 1958. Carbohydrate metabolism of mammary tissue, I. Pathways of glucose catabolism in the mammary gland. *Biochim. Biophys. Acta* 30:303.

Newsholme, E. A., and Gevers, W., 1967. Control of glycolysis and gluconeogenesis in liver and kidney cortex. *Vitamins and Hormones* 25:1.

Nordlie, R. C., and Arion, W. J., 1965. Liver microsomal glucose 6-phosphatase, inorganic pyrophosphatase, and pyrophosphate-glucose phosphotransferase. *J. Biol. Chem.* 240:2155.

Numa, S., Ringelman, E., and Lynen, F., 1965. Zur hemmung der Acetyl-CoA Carboxylase durch fettsaure Coaenzym A-verbindungen. *Biochem. Z.* 343:243.

Patton, S., 1974. Regulatory function of biological membranes. In *The Control of Metabolism* (J. D. Sink, ed.). The Pennsylvania State University Press, University Park, p. 75.

Pennington, R. J., 1952. The metabolism of short chain fatty acids in sheep. Fatty acid utilization and ketone production by rumen epithelium and other tissues. *Biochem J.* 51:251.

Potter, V. R., Baril, E. F., Watanabe, M., and Whittle, E. D., 1968. Systematic oscillations in metabolic functions in liver from rats adapted to controlled feeding schedules. *Fed. Proc.* 27:1238.

Ray, P. D., Foster, D. D., and Lardy, H. A., 1966. Paths of carbon in gluconeogenesis and lipogenesis. IV. Inhibition by L-tryptophan of hepatic gluconeogenesis at the level of phosphoenolpyruvate formation. *J. Biol. Chem.* 241:3904.

Riggs, A. D., and Bourgeois, S., 1968. On the assay, isolation and characterization of the lac repressor. *J. Mol. Biol.* 34:361.

Robinson, G. A., Butcher, R. W., and Sutherland, E. W., 1968. Cyclic AMP. *Ann. Rev. Biochem.* 37:149.

Schimke, R. T., 1962a. Adaptive characteristics of urea cycle enzymes in the rat. *J. Biol. Chem.* 237:459.

Schimke, R. T., 1962b. Differential effects of fasting and protein-free diets on levels of urea cycle enzymes in rat liver. *J. Biol. Chem.* 237:1921.

Schimke, R. T., 1963. Studies on factors affecting the levels of urea cycle enzymes in rat liver. *J. Biol. Chem.* 238:1012.

Schimke, R. T., 1969. On the roles of synthesis and degradation in regulation of enzyme levels in mammalian tissues. In *Current Topics in Cellular Regulation,* Vol. 1 (B. L. Horecker and E. R. Stadtman, eds.). Academic Press, New York, p. 77.

Schimke, R. T., and Doyle, D., 1970. Control of enzyme levels in animal tissues. *Ann. Rev. Biochem.* 39:929.

Schimke, R. T., Sweeney, E. W., and Berlin, C. M., 1965. The roles of synthesis of degradation in the control of rat liver tryptophan pyrrolase. *J. Biol. Chem.* 240:322.

Scrutton, M. C., and Utter, M. F., 1968. The regulation of glycolysis and gluconeogenesis in animal tissues. *Ann. Rev. Biochem.* 37:249.

Segal, H. L., 1959. Some consequences of the "non-competitive" inhibition by glucose of rat liver glucose-6-phosphatase. *J. Am. Chem. Soc.* 81:4047.

Shipley, R. A., Chudzik, E., Gibbons, A. P., Jongedyk, K., and Brummond, D. O., 1967. Rate of glucose transformation in the rat by whole-body analysis after glucose-C^{14}. *Am. J. Physiol.* 213:1149.

Srere, P. A., 1969. Some complexities of metabolic regulation. *Biochem. Med.* 3:61.

Stadtman, E. R., 1966. Allosteric regulation of enzyme activity. *Adv. Enzymol.* 28:41.

Tepperman, J., and Tepperman, H. M., 1970. Gluconeogenesis, lipogenesis and the Sherringtonian metaphor. *Fed. Proc.* 29:1284.

Tomkins, G. M., Gelehrter, T. D., Granner, D., Martin, D., Jr., Samuels, H. H., and Thompson, E. B., 1969. Control of specific gene expression in higher organisms. *Science* 166:1474.

Turkington, R. W., and Hill, R. L., 1969. Lactose synthetase:progesterone inhibition of the induction of α-lactalbumin. *Science* 163:1458.

Utter, M. F., and Scrutton, M. C., 1969. Pyruvate carboxylase. In *Current Topics in Cellular Regulation,* Vol. 1 (B. L. Horecker and E. R. Stadtman, eds.). Academic Press, New York, p. 253.

Vagelos, P. R., Alberts, A. W., and Martin, D. B., 1963. Studies on the mechanism of activation of acetyl CoA carboxylase by citrate. *J. Biol. Chem.* 238:533.

Veneziale, C. M., Walter, P., Kneer, N., and Lardy, H. A., 1967. Influence of L-tryptophan and its metabolites on gluconeogenesis in the isolated perfused liver. *Biochem.* 6:2129.

Villar-Palasi, C., and Larner, J., 1968. The hormonal regulation of glycogen metabolism in muscle. *Vitamins and Hormones* 26:65.

Villar-Palasi, C., and Larner, J., 1970. Glycogen metabolism and glycolytic enzymes. *Ann. Rev. Biochem.* 39:659.

Watkins, W. M., and Hassid, W. Z., 1962. The synthesis of lactose by particulate enzyme preparations from guinea pig and bovine mammary glands. *J. Biol. Chem.* 237:1432.

Wurtman, R. J., 1969. Time-dependent variations in amino acid metabolism: mechanism of the tyrosine transaminase rhythm in rat liver. *Adv. Enzyme Reg.* 7:57.

Subject Index

Author Index